Tornado F.2/F.3

Air Defence Variant

Hugh Harkins

Tornado F.2/F.3 - Air Defence Variant
© Hugh Harkins 2013

Published by Centurion Publishing
PO Box 3268
Glasgow
United Kingdom
G65 9YE

ISBN 10: 1-903630-38-X
ISBN 13: 978-1-903630-38-9

First published in 2004. This edition first published 2013

Cover design © Centurion Publishing & Createspace 2013

Page layout, concept and design © Centurion Publishing 2013

The publishers and author would like to thank the following for their assistance and contributions in the preparation of this publication: BAE Systems, MBDA, Panavia GmbH, RAF, Raytheon, Turbo Union GmbH and UK MoD

This No.43(F) squadron Tornado F.3 was painted in a special scheme in 1996 to celebrate the units 80th anniversary.
Author

Tornado F.2/F.3 - Air Defence Variant

Table of Contents

A pair of RAF Tornado F MK3's flies low across the desert floor during a deployment to the Gulf Region in March 2003, in support of the US, UK and Australian invasion of Iraq. Crown Copyright

Introduction

The Panavia Tornado ADV was adopted for the RAF to fight the predicted air battle over the North and Icelandic Seas. This could have involved large formations of Soviet long-range bombers, strike aircraft and maritime patrol aircraft, some of which would have been capable of releasing stand-off cruise missiles at distances of around 300-miles from the UK's shores. To meet the requirement, the RAF did not require a small agile fighter. The RAF required a long-range interceptor, which would be capable of carrying a load of at least eight air to air missiles and be capable of detecting the enemy at long-ranges and intercepting them as far from the UK as possible. In the 1970's, a number of off-the shelf options were looked at. A two-crew aircraft better served the demanding mission for the air defence of the UK and it was eventually decided to pursue a European solution by adopting a variant of the Tornado strike aircraft then in development. The resultant Tornado F MK3 was ideally suited to the long-range interception mission it was designed for.

The aircraft was plagued with protracted delays, particularly with the introduction of an acceptable capability standard of AI.24 Foxhunter radar. The early problems were overcome and the Foxhunter was developed into a highly capable radar system capable of detecting, tacking and engaging airborne targets at long-range. The introduction of a near real-time JTIDS secure datalink vastly increases the capability of the aircraft, which can share information with other airborne and surface based platforms.

While the Tornado F MK3 matured into a highly competent air defence interceptor, from the early 1990's, the aircraft has been deployed on operations in a counter-air role, a mission for which it is less well suited. However, despite reports to the contrary, the aircraft can still accomplish this mission competently. At the height of its service, the Tornado F MK3 equipped seven front line air defence squadrons, a Operational Conversion Unit, which was allocated a reserve squadron number plate, an operational evaluation unit, and an out of area flight based in the Falkland Islands. The Tornado ADV prototypes and a number of Tornado F MK2 and F MK3 production aircraft have also served with various test organisations.

New weapons and systems were introduced with the capability sustainment program initiated in the late 1990's. This along with other upgrade programs will allow the aircraft to remain in service until its planned out of service date of 2010, when it is scheduled replacement by the Eurofighter Typhoon is complete. Introduction of active-radar guided AMRAAM and the highly capable infrared guided ASRAAM combined with radar upgrades have increased the capability of the aircraft in an interceptor, counter air fighter capacity. The ASRAAM is a quantum leap in capability over the AIM-9L Sidewinder it replaced allowing engagements of enemy aircraft in a beyond visual range arena with infrared guided missiles. From early 2003, the Tornado F MK3 at last began to demonstrate some of the true multi-role potential of the design when aircraft serving with No.11 squadron at Leeming took on a suppression of enemy air defence role armed with the MBDA ALARM 2 air to surface anti-radiation missile.

August 14, 2004 marked the 30th anniversary of the first flight of the first Tornado prototype. The 30th anniversary of the first flight of the first Tornado ADV prototype was marked on 27 October 2009 and the 25th anniversary of the first flight of a production Tornado F Mk.2 was marked on 5 March 2009. The 25th anniversary of the first deliveries to the RAF was marked on 5 November 2009. The 25th anniversary of the official entry into service and first flight of the Tornado F Mk.3 was marked in 2009, with the last Squadron disbanding in March 2011 after 27 years service with the RAF.

This volume covers the genesis of the Tornado program and the emergence of the ADV fighter variant. The Tornado ADV is described in detail, as is its operational service with the RAF, Saudi Arabia and Italy. A chapter looks at the upgrades to mission systems and weapons including air to surface weapons allowing the Tornado ADV to remain at the cutting edge of air defence technology into the second decade of 21st century when it was retired from RAF service.

Chapter One

Tornado Genesis

The Panavia Tornado F.3 air defence interceptor was developed from the MRCA (Multi Role Combat Aircraft) design, which emerged as the Tornado IDS (Interdictor Strike). Although under powered at medium and high altitudes, the Tornado F.3 variant in service with the RAF has matured into a potent air defence fighter. Upgraded with new systems and weapons, the Tornado F.3 is scheduled to serve with the RAF until 2010, by which time it should have been completely replaced by Eurofighter Typhoon
Crown Copyright

Like most modern combat aircraft, the origins of the Tri-National Panavia Tornado can be traced back through a labyrinth of false starts and cancelled programs over a number of decades. The Tornado program, which was originally known under the title of Multi Role Combat Aircraft (MRCA), was the result of the search for a replacement for a number of aircraft types in several European nations.

The hugely controversial British Aircraft Corporation (BAC) TSR-2 strike aircraft program for the RAF still stirs emotions almost forty years after its final demise. This aircraft was in many respects the most advanced strike aircraft design of its time. Had it gone ahead, it is possible that variants of the TSR-2 design would still be serving with the RAF today. The TSR-2 program fate was sealed when its cancellation was announced in the Labour government budget speech on 6 April 1965.

Following the demise of the TSR-2, the United Kingdom and France entered into an agreement to design and develop a strike aircraft known as the AFVG (Anglo-French Variable-Geometry) on 17 May 1965.

Had the BAC TSR-2 strike aircraft program not met an untimely end in 1965, the UK would have produced the aircraft in a number of variants, which could have included a long-range interceptor. Crown Copyright

BAC put forward the BAC P.45 variable-geometry design to meet the AFVG requirement. In AFVG guise, the P.45 was to have been powered by the French M45 engine. The troubled AFVG fell by the wayside when France unilaterally withdrew from the program on 5 July 1967, leaving the UK to re-commence its search for a future strike aircraft program. Following the French withdrawal from the AFVG program, the UK looked at powering the P.45 with the Rolls Royce RB.153 from the cancelled TSR-2. Had the P.45 design progressed it would probably have been fitted with much of the avionics developed for the TSR-2.

The P.45 design was eventually dropped as the UK wanted to proceed with a collaborative European combat aircraft program, which made sense from a development risk and cost point of view. However, as time has proven, collaborative European combat aircraft programs come fraught with not only technical difficulties, but political differences which often delay and compromise the program, although the risk sharing and development cost benefits are obvious.

France continued development of a variable geometry combat aircraft known as the Dassault

Mirage IIIG, which was developed from the hugely successful Dassault Mirage III fighter. This single engine aircraft conducted its first flight on 18 November 1967. It was intended that this interim aircraft would pave the way for a carrier variant known as the Mirage IIIG2 and a twin-engine variant known as the Mirage IIIG4. These variants were cancelled on cost grounds. However, a pair of Mirage IIIG8 prototypes was built under the ACF (*Avion de Combat Futur*) program. These aircraft were reduced cost twin-engine variants of the Mirage IIIG. The 2 Mirage IIIG8's conducted their first flights on 8 May 1971 and 3 July 1972 respectively. However, the Mirage IIIG8 program was cancelled in 1975, with France going on to produce the Mirage 2000 family of combat aircraft.

Above: Initially called the MRCA (Multi-Role Combat Aircraft), the first Tornado prototype, P01 (D-9591) conducted its first flight from MBB (Messerscmitt-Bolkow-Blohm) Manching facility in the then West Germany on 14 August 1974. EADS

In the 1960's, West Germany was searching for a future strike aircraft to replace its large fleet of Lockheed F-104G Starfighters. The AVS (Advanced Vertical Strike) was an unpractical program, which resulted in a Vertical/Short Take-Off and Landing (V/STOL) combat aircraft design jointly developed by the USA and West Germany. Republic Aircraft in the USA had teamed with Fokker of the Netherlands to submit the D.24 in the 1961 NATO NBMR-3 competition for a V/STOL strike fighter. The pivoted variable-geometry D-24 design was to be powered by a single 120-kN (27,000-lb) Bristol Siddeley BS.100 lift/cruise engine. In addition, separate lift-engines would be provided for vertical take-off and landing. This complicated and completely unpractical design, like so many before it, was cancelled.

Top: The second British Tornado prototype, P03 (XX947) is seen during a test flight in the 1970's. Above: P05 (MMX856) was the first Italian Tornado prototype to fly. BAE Systems and Alenia

Around the same time, Boeing was working with EWR on the EWR.360, which like the D.24 featured pivoted swing-wings. When Boeing dropped out of the program, Republic (renamed Fairchild Republic in 1965) took its place as part of the AVS program. This design would have employed a new lift-jet designed by Rolls Royce and Allison and designated XJ99.

With the collapse of the AFVG, Britain was left without a future strike aircraft program other than the Anglo-French SEPECAT Jaguar and the Blackburn Buccaneer S.2. At the time, the Jaguar was being designed as a twin-seat advanced supersonic trainer and as a single-seat day fighter-bomber. The Buccaneer was not favoured by the RAF because it was a naval design, although the

Buccaneer S.2 would eventually be procured and go on to serve the RAF until 1993.

In July 1968, the UK joined the multinational MRCA (Multi-Role Combat Aircraft program). This program emerged in January 1968, when Belgium, Italy, The Netherlands and West Germany embarked upon a program to replace their respective fleets of Lockheed F-104 Starfighter's. A joint working group was commissioned and this commenced work on 3 March 1968, by which time Canada had joined the program with replacement of its Starfighter's in mind. The UK joined the program in July as the only non-Starfighter operator. On 17 July 1968, an agreement was reached between the six partner nations for the launch of the concept development phase of the program, which continued until 30 April 1969.

Top: This artist impression of the Tornado ADV was released by Panavia in 1978. It clearly shows the sleeker lines of the ADV, with the Sky Flash missiles semi-recessed in staggered formation under the fuselage. Above: A Tornado ADV flies in formation with the two aircraft types it was procured to replace, the McDonnell Douglas Phantom FGR.2 and BAe Lightning. Panavia GmbH and BAE Systems

The first of the three Tornado ADV prototypes to fly was A01 (ZA254), which conducted its maiden flight on 27 October 1979. Early test flights demonstrated that the ADV possessed better supersonic acceleration than the Tornado IDS variant due to the improved fineness ratio of the fuselage. The more forward centre of gravity caused the aircraft to demand greater elevator angle at take-off than that required by the IDS variant. BAE Systems

During the concept development phase, it was agreed to develop both single and twin-seat variants of the MRCA (later known as the Tornado 100 and Tornado 200 respectively). However, this was later abandoned in favour of all twin-seat variants, as two-crew were deemed more suitable for the high workload involved in flying an operational mission, particularly at low-altitude.

Canada and Belgium withdrew from the program in late 1968, leaving the other four partner nations to form Panavia Aircraft GmbH on 26 March 1969. This new company was headquartered in Munich, West Germany. A single-organisation was also formed for submission of partner nation requirements to Panavia. This was known as NAMMA (NATO Multi-Role Combat Aircraft Management Agency), which was formed on 15 December 1968. This agency is now part of the NATO Eurofighter Tornado Management Agency (NETMA). The Netherlands left the program in July 1969, leaving the other three partner nations to thrash out work share and production planning. Work share was allocated in accordance with planned procurement for each of the three partner nations. This was set at 42.5% each for the UK and West Germany and 15% for Italy.

The new fighter was to be powered by an all-new afterburing turbofan engine. A Rolls Royce engine was chosen as the basis of the new engine, which became known as the RB199. Rolls Royce (UK), MTU (West Germany) and Fiat (Italy) formed a new company called TurboUnion GmbH to produce the new engine. Avionics for the MRCA were to be developed by a new company called Avionics Systems GmbH. While the new aircraft and engine were being designed and built as a European solution to future strike aircraft requirements, the MRCA was let down by selection of a US developed navigation and attack radar. This system, developed by Rockwell was known as the TNR (Tornado Nose Radar).

The first Tornado ADV, A01 (ZA254) is seen with wings in the 67-degree sweep position. The aircraft is carrying dummy Skyflash missiles on the four under fuselage stations and a pair of IDS 1500-litre (330-gallon) external fuel tanks on the wing stations. The attractive black and while livery, which adorned the first two ADV prototypes contrasts sharply with the dark background of the Irish Sea. BAE Systems

The preferred solution to the challenging design for strike aircraft designed to operate at ultra-low-level in foul weather day or night in an extremely hostile Electronic Counter Measures environment was a variable Geometry (swing-wing) design. Aircraft design has generally followed trends, which can be likened to fashion. In the late 1960's and early 1970's, variable geometry was seen as the best approach to getting the best performance from an aircraft in multiple flight regimes. Most major combat aircraft building nations were developing variable geometry designs at the time. The US was producing the General Dynamics F-111 strike aircraft, the Soviet Union was developing the Sukhoi Su-22 'Fitter' and Su-24 'Fencer' strike fighters and the MiG-23/27 family of fighters. The US also chose a variable geometry design as its new naval fighter in the early 1970's, which emerged as the Grumman F-14 Tomcat. On a larger scale, the US and Soviet Union introduced variable geometry strategic bombers into service in the 1980's, the Rockwell B-1B Lancer and the Tupolev Tu-160 'Blackjack' respectively. The later has the distinction of being the heaviest combat aircraft ever to enter service. The Soviet Union also developed the Tu-22 'Backfire' variable-geometry intermediate range bomber. The variable geometry wing design would allow the MRCA to emerge as arguably the best low-level strike aircraft of its time. Far outperforming its rivals and the aircraft it was designed to replace.

The Tornado was designed to fill seven different roles for the air forces of the three partner nations and the naval air arm of the then West Germany. The roles for which the type was designed were **1:** Interdiction strike (penetrating to attack targets deep inside enemy territory) **2:** Counter air attacks (attacks on enemy airfields) **3:** Battle field air interdiction. **4:** Close air support. **5:** Reconnaissance. **6:** Maritime strike **7:** Point interception. This last mentioned role was not adopted for any of the Tornado IDS operators, although the British RAF adopted the Tornado Air Defence Variant to fulfil the long range all weather day night interception-role to meet AST.395, with preliminary studies commencing around 1969.

Wings in the fully swept back position, the second Tornado ADV prototype, A02 (ZA267) demonstrates its sleek lines in the head-on aspect. A02 conducted its first flight on 18 July 1980. BAE Systems

Once it entered service, another role was allocated to the Tornado, suppression of enemy air defenses with both Italy and Germany procuring the purpose designed ECR (Electronic Combat and Reconnaissance) variant armed with the US AGM-88 HARM (High Speed Anti Radiation Missile). Britain went the other road and adopted the ALARM (Air Launched Anti-Radiation Missile) for its standard Tornado GR.1s.

A tri-nation Memorandum of Understanding (MoU) signed on 22 July 1970 authorised full-scale development of the MRCA. The MoU included building 10 prototypes, which today would be referred to as development aircraft. Nine of these were to be used for flight-testing and the tenth was a static test airframe. P01, P04 and P07 were assembled at Manching, West Germany. BAC

(later BAe) at Warton in the UK assembled P02, P03, P06 and P08. An additional airframe, P10 was built as the static test airframe at Warton. P05 and P09 were built in Italy. The Tornado was the first aircraft in the world designed with a FBW (fly-by-wire) flight control system. However, it was beaten into the air by the less complex General Dynamics YF-16A prototype, which conducted its first official flight at Fort Worth, Texas on 2 February 1974. The first Tornado (at this time still known as MRCA) prototype, the West German assembled P.01 (D-9591) was rolled out at Manching on 8 April 1974. P01 took to the air for the first time on 14 August 1974, with BAC's Paul Millet at the controls. On 31 October 1974, the first British aircraft, P0.2 (XX946), which had been called Tornado in September 1974, conducted its maiden flight from Warton. P.03 (XX947) conducted its first flight on 5 August 1975, followed by the second West German aircraft, P04 (D-9592), which conducted its first flight on 2 September 1975.

The last of the three Tornado ADV development aircraft was A03 (ZA283), which conducted its maiden flight on 18 November 1980. Visible on the vertical tail leading edge is the camera fairing used to record elements of the flight test program. BAE Systems

The first Italian assembled Tornado; P05 (MMX586) conducted its first flight on 5 December 1975. P06, the third British assembled Tornado, conducted its first flight on 19 December 1975. The West German assembled P07 (98+06) flew for the first time on 30 March 1976, followed by P08 (XX950), the fourth British assembled Tornado, which flew at Warton on 15 July 1976. The Italian assembled P09 (MMX587) was the last of the flying prototype Tornado's to fly when it took to the air for the first time on 5 February 1977. Of the nine Tornado prototypes P08 was lost in an accident in June 1979 and P08 was lost in an accident in April 1980.

The Nine flying prototypes were joined by six pre-production Tornados'. The first of these to fly was the West German P11 (98+01), which first flew on 5 February 1977, the same day P09, the last of the nine flying prototypes conducted its first flight in Italy. Just over one month later, the British assembled P12 (XZ630) conducted its first flight when it lifted off from Warton on 14 March 1977. The West German P13 (98+02) flew for the first time on 10 January 1978. The Italian pre-production aircraft, P14 (MMX588) conducted its maiden flight on 10 January 1979. The last of the

pre-production Tornado IDS, the British assembled P15 (XZ631) had flown at Warton on 24 November 1978.

With the MRCA development program well underway, the UK turned its attention to a variant of the new combat aircraft, which could replace the BAC Lighting and McDonnell Douglas Phantom FGR.2, and eventually the Phantom FG.1 in service with RAF interceptor squadrons. The new variant would eventually emerge as the Tornado Air Defence Variant (ADV), which was initially a much more low profile program than that of its strike cousin. Preliminary studies of the interceptor variant began in 1969, with full-scale development of the ADV being launched in 1976.

While in development, the fighter variant of the Tornado promised to be amongst the best interceptors in the air defence world. It was anticipated that the new fighter would be adopted by a number of European NATO nations, in particular the two remaining partner nations, Italy and West Germany. However, neither nation was looking to replace their fleets of air defence fighters, the Lockheed F-104S in Italy or the McDonnell Douglas F-4F in West Germany. Both countries would embark upon life extension upgrade programs for their respective fighter fleets, shunning the Tornado ADV. Former MRCA partners Belgium and The Netherlands had both purchased the General Dynamics (now Lockheed Martin) F-16/AB, which was a much smaller cheaper alternative to either Tornado variants. While the smaller single-engine F-16 would be

Top: The Tornado ADV prototype (background) conducts a formation take off demonstration with a Tornado GR.1 IDS variant. Above: Complete with dummy Skyflash and Sidewinder air to air missiles, the first prototype ADV, ZA254 has the wings in the forward sweep position during a demonstration flight in 1985. Both US DoD

The second Tornado Air Defence Variant development aircraft A02, ZA267 was used to demonstrate the air to surface and to a lesser degree some of the multi-role potential of the design. The aircraft is seen carrying a pair of Kormoran anti-ship missiles on fuselage stations, a pair of Skyshadow electronic countermeasures pods converted to carry test equipment on the outer wing stations, a pair of 1500-litre external fuel tanks on the inboard wing stations with a pair of AIM-9L Sidewinder infrared guided air to air missiles on the inboard wing station stub pylons. This configuration was used to try and interest the Japan Air-Self Defence Force, which was looking for a new support fighter. BAE Systems

developed into a versatile affordable multi-role combat aircraft, it fell short of the Tornado IDS capabilities as a strike aircraft or the Tornado ADV's capabilities as an interceptor. The other former MRCA partner, Canada, eventually chose the McDonnell Douglas CF-18 Hornet to replace its fleet of Starfighter's and McDonnell Douglas CF-101 Voodoo interceptors

With the Tornado IDS development program progressing well, the Tornado Interdictor Strike (IDS) was ordered into production for the three partner nations. Included in Batch One of Tornado production were three prototypes Tornado ADV's for the UK. The Tornado ADV benefited from much of the development work conducted for the IDS variant such as basic flight-testing, FBW flight-control development and 27-mm Mauser cannon testing. The large 2250-litre 'Hindenburger' external fuel tank developed for the ADV was flight-tested on a Tornado IDS.

The first two ADV prototypes were painted in an attractive black, white and grey colour scheme, which made the aircraft highly conspicuous. The aircraft were also covered in black crosses to assist in analysing high-speed photography. The First development aircraft, A01 (ZA254) was rolled-out at Warton on 9 August 1979 and conducted its first flight at Warton on 27 October 1979, with David Eagles at the controls and Roy Kenward in the back seat. Even at this early stage the aircraft was

A Tornado ADV development aircraft employs full afterburner with the wings swept in the 67-degree position for high speed flight. US DoD

carrying four dummy Skyflash missiles. On this first flight, which lasted 22 minutes, the aircraft exceeded Mach 1 and before a week was out had flown 8.5 hours in five flights, which included an air to air refuelling and a night landing. These early flights had proved that the ADV had better supersonic acceleration than the Tornado IDS variant due to the improved fineness ratio of the fuselage. However, the more forward centre of gravity caused the aircraft to demand greater elevator angle at take-off than that required by the IDS variant.

The second development Tornado ADV, A02 (ZA267), conducted its maiden flight on 18 July 1980. This was a dual-control aircraft and was assigned to weapons development. The aircraft incorporated improvements including a main computer and associated cockpit TV displays. The third and last of the three ADV development aircraft, A03 (ZA283), was flown on 18 November 1980. This aircraft had an overall grey colour scheme similar to that adopted by the RAF for service Tornado F MK2/3's. This aircraft was to be used for radar development, but due to delays with the A1.24 Foxhunter radar did not begin radar trials until June 1981.

The prototype ADV's had a gantry installed at the base of the fin, which housed a large parachute. In the event of the aircraft entering a spin in which recovery action failed to work the parachute would be streamed. This rapidly lowers the aircraft's nose, reducing the incidence and stopping the spin. Internally, the aircraft had an emergency power unit driven by a large hydrazine pump. This provided hydraulic and electrical power to drive the control system in the event of a double engine flameout during a spin. This had a back up in the

form of an Electro-hydraulic pump, which does the same thing. The prototypes carried inert BAe Dynamics Skyflash semi-active medium range air to air missiles, which were painted yellow and had black test crosses for high-speed photography of the launch sequence. To photograph the launch, the Tornado was equipped with a pair of camera pods in the shape of converted Skyshadow ECM pods as carried by the Tornado IDS.

By mid-1980, the first prototype A01 (ZA254) had flown at an Indicated Air Speed (IAS) of 800-kts (921-mph, 1480-km/h) at a height of 610-m (2,000-ft). This capability would give the ADV a considerable speed margin over most of its potential adversaries, as most modern combat aircraft are limited to about 700-750-kts due to structural design limitations. This meant that nothing could escape from a Tornado ADV when being pursued at low altitude. In early 1982, the same aircraft demonstrated its ability to fly a CAP (Combat Air Patrol) of two hours and twenty minutes at a distance of 375-nm from base. The aircraft achieved this without aerial refuelling by using 1500-litre (330-gal) external fuel tanks as used on the Tornado GR.1 (IDS) strike aircraft. When the Tornado F MK3 entered service with its more fuel-efficient RB199 MK104 turbofans and 2250-litre (495-gal) external fuel tanks, performance was increased further. To top it off, on its arrival at Warton A01 loitered in the airfield's vicinity for 15 minutes before landing with more than 5% fuel reserves remaining.

In the same year, A02 carried out various armament trials involving firing Skyflash missiles at M0.9 up to supersonic speeds. The internal Mauser 27-mm cannon located on the starboard forward fuselage was also tested in the subsonic flight regime above 200-kts from zero-g up to the angle of attack limit and from low-level up to 30,000-ft.

Chapter Two

Tornado Air Defence Variant Described

Above: A Tornado F.3 in No.111(F) squadron's 85ᵗʰ anniversary scheme is seen carrying a load of four MBDA ASRAAM missiles in 2002. MBDA/G H Lee

Designed in the late 1960's and early 1970's, the Panavia Tornado was among the last generation of combat aircraft to be built before the large-scale use of carbon-fibre composite materials was used in combat aircraft construction. The fuselage of the Tornado is of all metal, semi-monocoque construction, and is built in three separate sections. The front fuselage and rear fuselage was built by BAe. The upper rear fuselage has two large door type airbrakes that are extended by hydraulic jacks. They form part of the skin on each side of the upper rear fuselage. The centre section, including

A No.43(F) squadron Tornado F.3 shows its underside revealing the main armament of four MBDA Skyflash air-to-air missiles. Crown Copyright

the engine air intakes and wing pivoting mechanism, is built by MBB. The fuselage houses most of the aircraft's fuel in Uniroyal, multi-cell self-sealing integral tanks. Other fuel tanks are located in the wings and the tail fin. Refuelling is accomplished through a NATO standard single-point pressure connector. A fuel jettison pipe is located on the fin trailing edge above the rudder. A built-in arrestor hook is located in the rear fuselage below the engine bays, which assists if the aircraft is forced to land within a very short distance, for example a damaged runway. The landing gear consists of a hydraulically retractable tricycle unit, with Dunlop single-unit main wheels and a twin-nose wheel with multi-disc brakes, which carry low-pressure heavy-duty tyres and Goodyear anti-skid units allowing operations from semi-prepared surfaces.

The variable-geometry wings are of all-metal construction. The fixed inboard section has a 60-degree sweep back on the IDS variant and 67-degree of sweep back on the ADV. The outer movable section can be swept from the 25-degree fully forward position to the 67-degree fully swept position. The wing has no ailerons, but has two spoilers in the upper surface forward of the trailing edge, which augment roll control at intermediate and fully forward wing sweep.

No.56(Reserve) squadron Tornado F.3 ZH556 with wings in the fully swept back position. Crown Copyright

The Tornado ADV (background) was developed from the Tornado IDS (foreground), which emerged from the initial MRCA design of the early 1970's. There is little doubt that the Tornado F.2/3 design as a pure fighter was compromised by the adoption of a design optimised for the low-level strike mission. However, despite its strike aircraft design lineage, the Tornado F.3 emerged as a competent all-weather fighter especially in the beyond visual range arena. BAE Systems

The entire outer edge of the wing trailing-edge, with the exception of the wing tips, incorporates full span, fixed vein, double slotted fowler flaps in four sections, while the wing leading edge has three slats. The wings of ADV and the IDS are basically the same; however, the F.3 variant of the ADV has an auto wing sweep facility with four pre-programmed settings. 25-degree fully forward for speeds up to Mach 0.73, 45-degree for speeds up to Mach 0.88, 58-degree for speeds up to Mach 0.95 and the fully swept position of 67 degree for speeds up to Mach 2.2.

The swept tail unit is of cantilever all-metal construction and comprises a single broad chord, swept twin spar vertical tail fin, with rudder mounting with low set all moving horizontal surfaces called tailerons. The tailerons and rudder are actuated by electrically controlled hydraulic jacks, operating in unison for pitch control or differentially for roll control.

All Tornado variants are carried by singe-unit main wheels, which retract into the fuselage and a forward retracting twin-nose wheel unit. This aircraft is about to be lifted by a crane and dropped to demonstrate the high impact absorption of the undercarriage units. Author

General-arrangement drawing of the Panavia Tornado F Mk.3. Crown Copyright

The one-piece cockpit canopy opens upwards and is hinged at the rear. The steeply raked windscreen built by Lucas Aerospace comprises a flat centre panel and two curved side panels, which are heavily armoured in order to meet the RAF's strict bird-strike requirements. Sierracotte electrical conductive heating is incorporated in the entire windscreen. This is effective for anti-icing and de-misting. The windscreen is kept free of rain by rain dispersal ducts, which bleed hot pressurised air from the aircraft's air-conditioning system. The whole windscreen area hinges forward, allowing access to the pilot's instrument panel and the HUD (Heads Up Display). Beneath the canopy the two-crew sit on Martin Baker MK10A zero-zero ejection seats. In the case of ejection an MDC (Miniature Detonating Chord) in the canopy detonates to shatter the canopy a split second before the seat fires.

The main changes between the Tornado IDS and ADV are internally with the aircraft sharing 80% commonality with each other. Among the external changes was the ADV having a 1.36-m (4-ft 5.5-in) fuselage extension, which resulted in the ADV having a more pointed nose profile compared with the IDS. This allowed an extra bay forward of the wings, and also allowed the aircraft to carry four BAe Dynamics (now Matra BAe Dynamics Alenia) Skyflash SARH (Semi-Active Radar Homing) MRAAM (Medium Range Air-to-Air Missiles) under its fuselage, with the front pair being semi-recessed. Another of the offsets was the ability to carry additional internal fuel extending the Tornado's already impressive range.

A Tornado F.3 head-on at low altitude. Crown Copyright

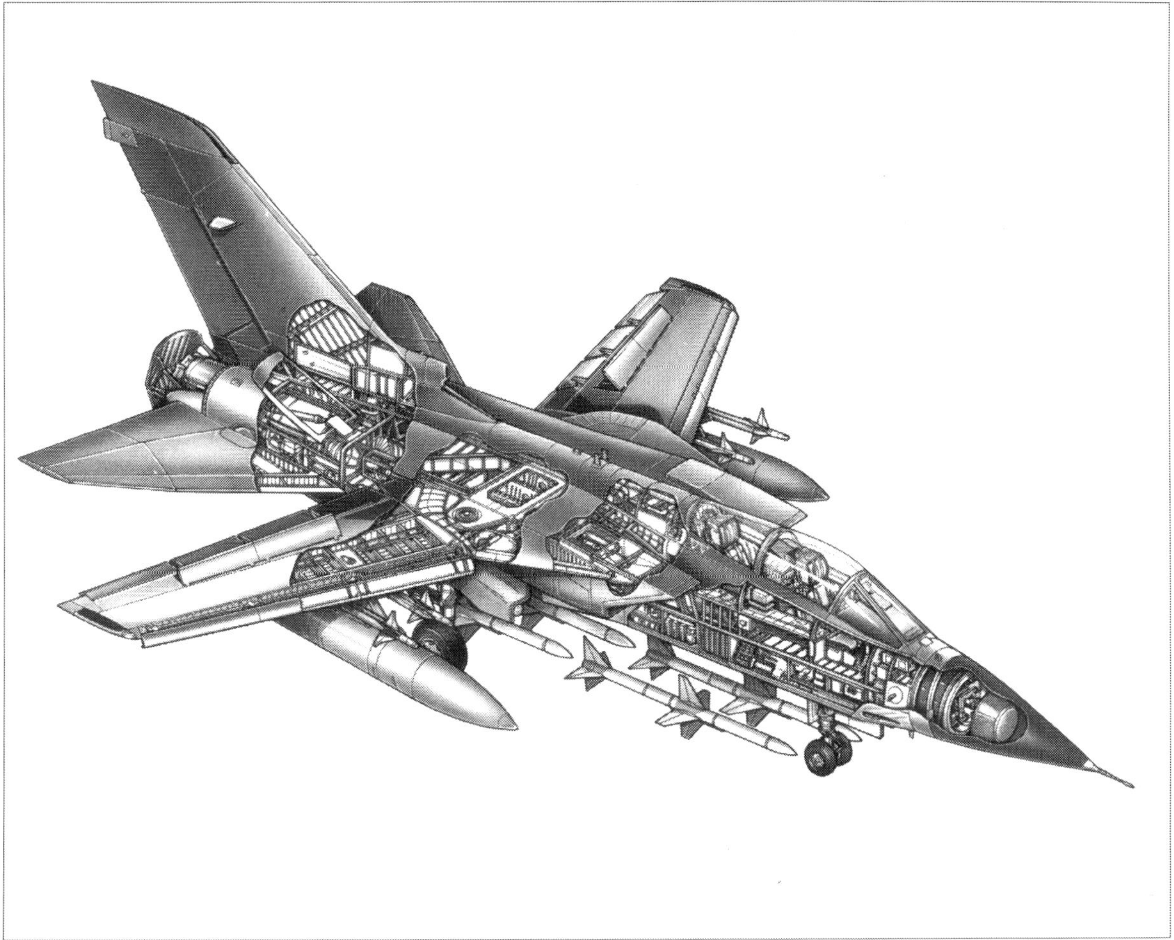

Semi-cutaway view of the Tornado ADV showing radar, wing-mechanism, armament and engines including deployed thrust-reverse buckets. BAE Systems

In addition to its extra internal fuel compared with the IDS, the F.3 can also carry the much larger 2250-litre (495 imperial gallon) 'Hindenburger' external fuel tank, one carried on each of the inboard under wing stations. Following the experience gained during the 1991 Gulf War, Tornado GR.1/4 squadrons regularly fly equipped with the larger external fuel tanks. The inboard stations of the F.3 have been fitted with stub pylons able to carry an AIM-9L/M or a MBDA ASRAAM (Advanced Short-Range Air to Air Missile) on each side of the fuel tank.

The Tornado ADV was also equipped with a second Inertial Navigation System (INS) to compensate for the loss of the Doppler velocity sensing radar used on the Tornado IDS for basic navigation information. The INS is autonomous in operation in that it requires no external inputs apart from feeding in the aircraft's precise position at the start of each flight. However, the system suffers from slippage, creating positional errors over say a two-hour sortie. Therefore, the Tornado crew must update the system with fixes throughout the flight.

A Tornado F.3 banks to port with wings in the maximum sweep position. The aircraft is carrying a load of four Skyflash and four Sidewinder air-to-air missiles, but lacking external fuel tanks. Crown Copyright

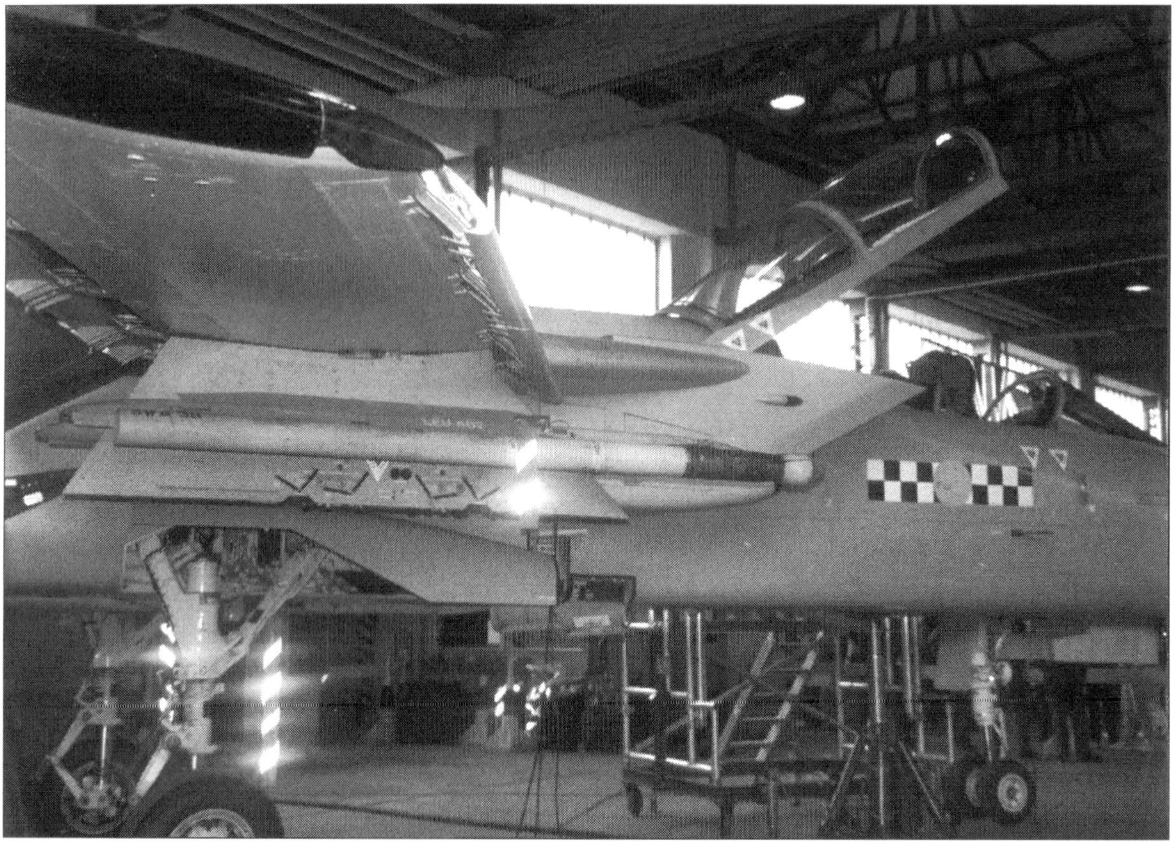

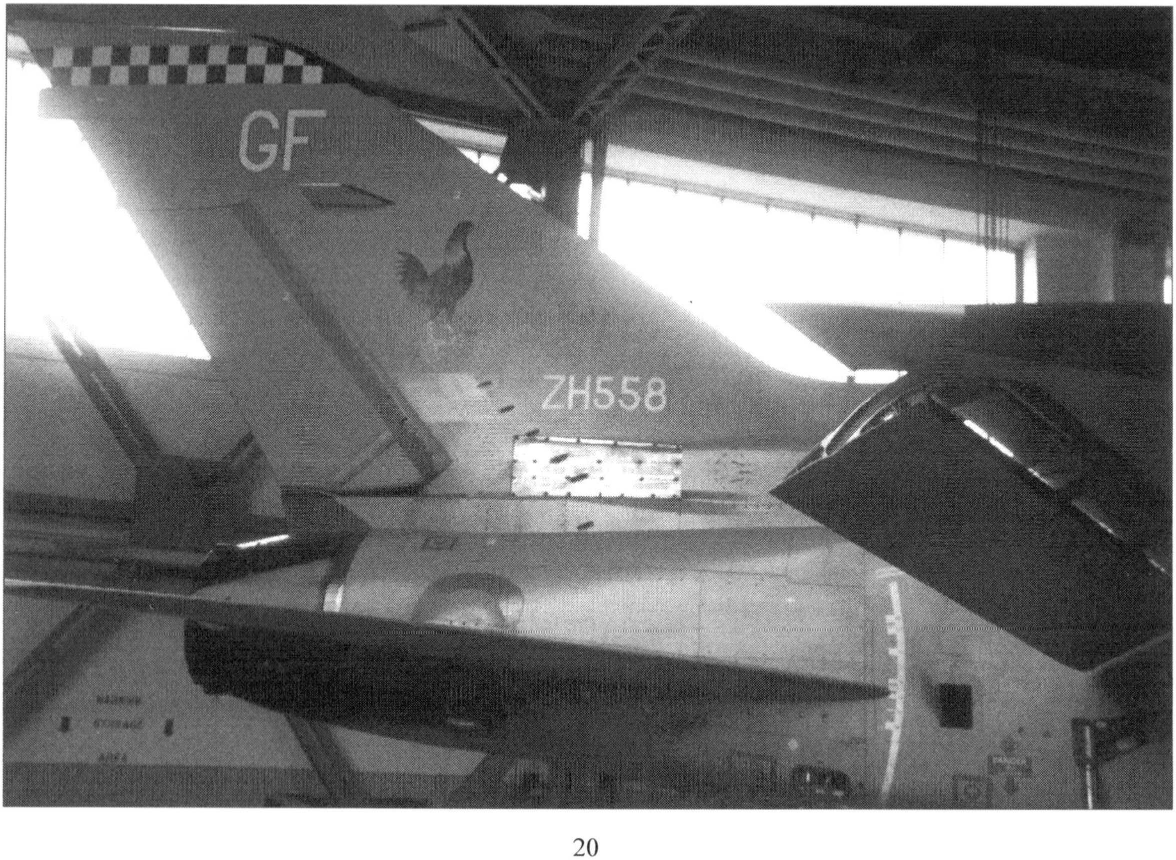
GF
ZH558

Previous page top and bottom: The variable-geometry wings have no ailerons, but have two spoilers in the upper surface forward of the trailing edge, which augment roll control at intermediate and fully forward wing sweep. The entire outer edge of the wing trailing-edge, with the exception of the wing tips, incorporates full span, fixed vein, double slotted fowler flaps in four sections, while the wing leading edge has three slats. This page top: The Tornado F.3 is equipped with the large 2250-litre external fuel tank, which together with the port side fully retractable in-flight refuelling probe, significantly extends the combat radius of the aircraft. All Author

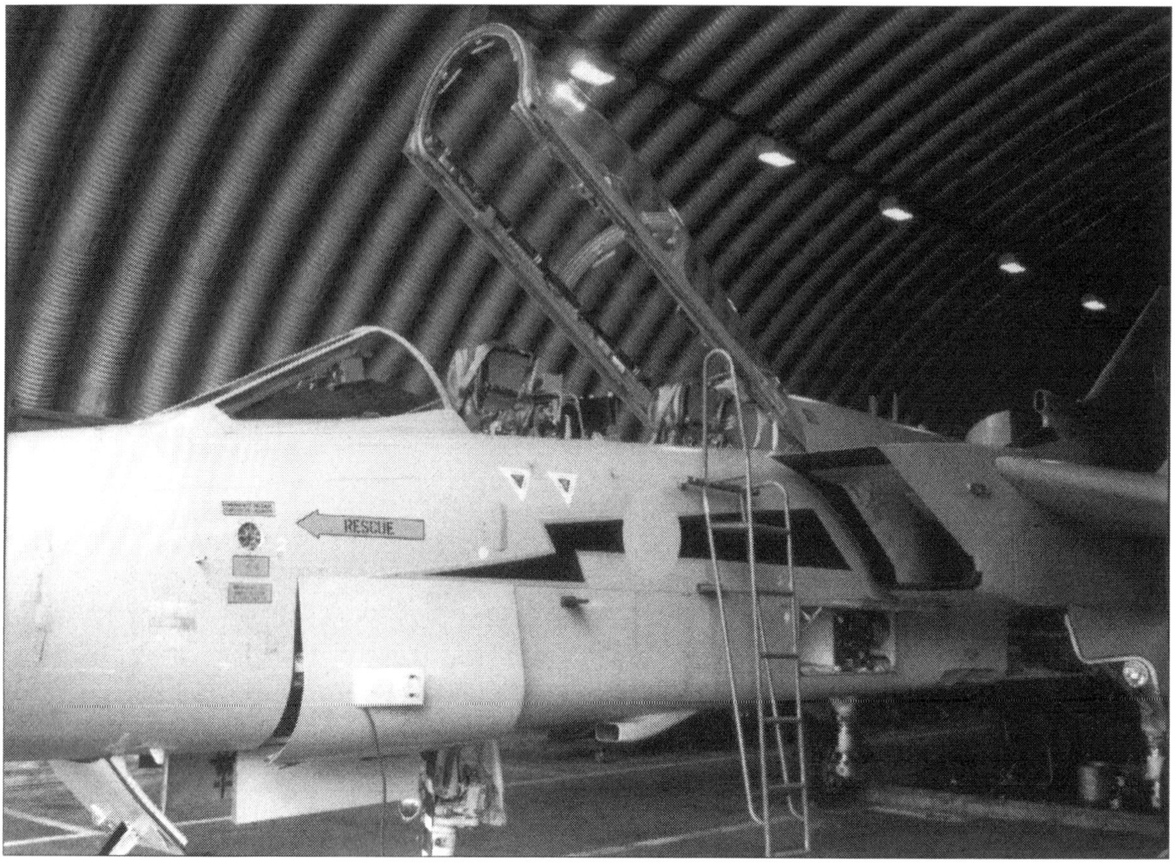
RESCUE

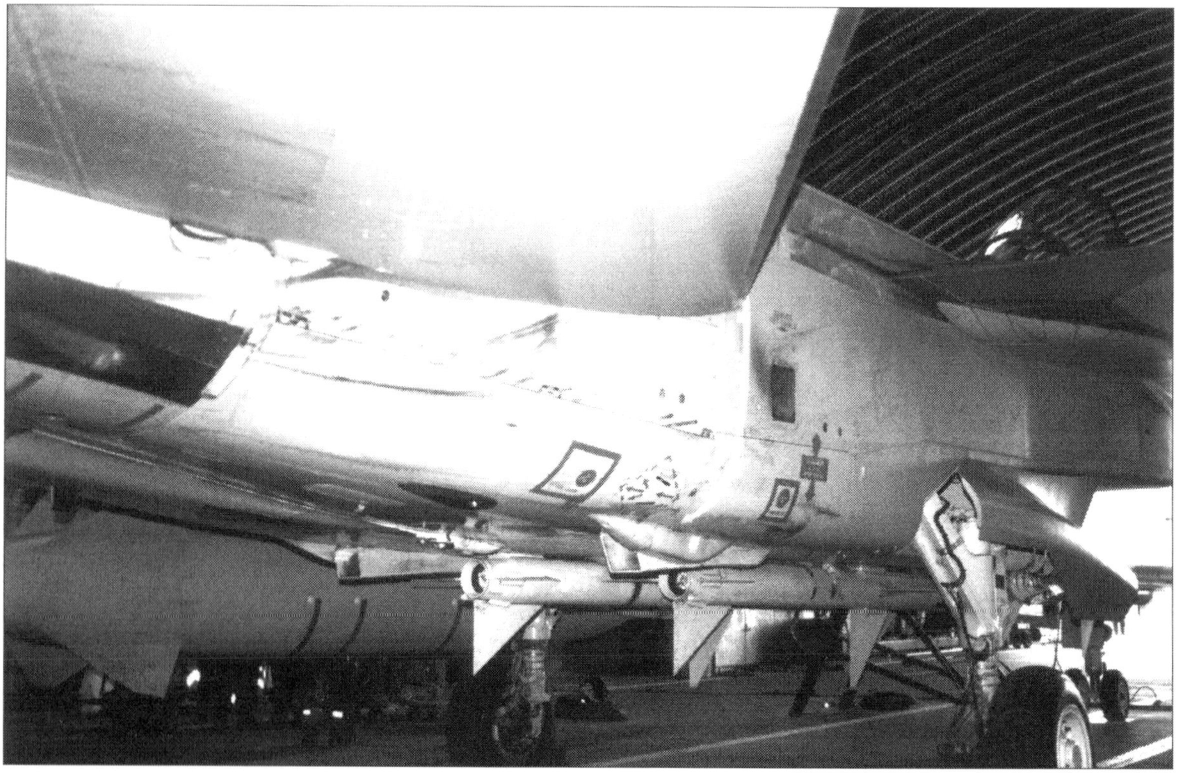

Previous page top: A No.111(F) squadron Tornado F.3 is seen in a HAS (Hardened Aircraft Shelter) at Leuchars with cockpit canopy open and BITE (Built In Test Equipment) equipment hatches open. Previous page top: Like the Phantoms it replaced, the Tornado F.3 carried its main armament of four Skyflash BVR missiles under the fuselage. Just aft of the rear two missiles can be seen the empty flare dispenser boxes which are scabbed on when the aircraft is on operations. Both Author

This page top and centre top left: The Tornado was designed with ease of maintenance, Incorporating a number of BITE hatches for easy location, analysis and fixing of problems. Centre bottom left: This view from the spine of a Tornado F.3 shows the exhaust blast vents located at the rear of the HAS allowing full engine running in the shelter prior to taxiing. Bottom left: The Tornado is equipped with two airbrakes located on either side of the vertical tail. The brakes are extended in this view of ZE250 undergoing maintenance at Leuchars. The Tornado F.3 can carry the large 2250-litre 'Hinderburger' external fuel tanks top right, or the smaller 1550-litre external fuel tank bottom right. One tank can normally be carried on each of the inboard wing stations. All Author

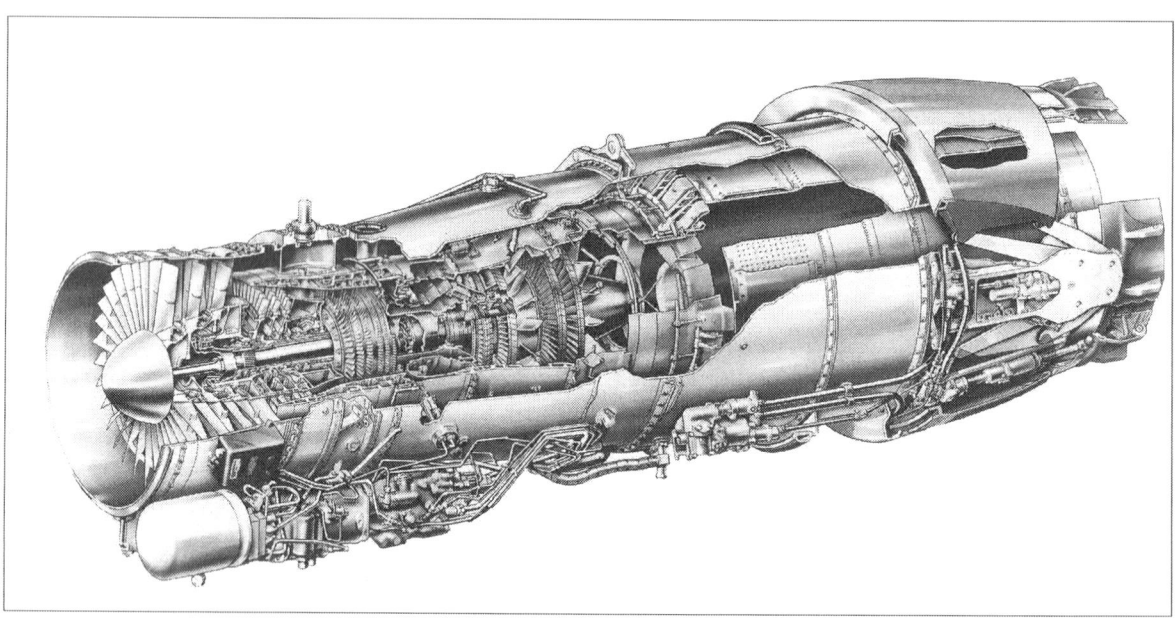

Top: One of the main differences between the RB199 MK104 and the RB199 MK103 was the former having a 14-inch extension to the afterburner area seen on Tornado F.3 ZE835. Author

Above: A cutaway diagram of the RB199 showing internal workings including the front-end fan and rear-end afterburner section. Rolls Royce

At medium to high altitudes the Tornado is under-powered, particularly for the counter air fighter role. However, at lower altitudes, the RB199 comes into its own powering the Tornado at speeds up to Mach 1.2 (800-kts). This No.11 squadron Tornado F.3 breaks left showing its under fuselage load of two MBDA ALARM 2, four MBDA ASRAAM (Advanced Short-Range Air to Air Missile) on the inboard stub pylons a pair of external 1500-litre fuel tanks on the inboard stations and a phimat chaff dispenser pod and Ariel towed decoy on the right and left outboard stations respectively. G H Lee

While the Tornado IDS was equipped with two internal Mauser 27-mm cannon on the lower forward fuselage, the Tornado ADV is equipped with only a single cannon on the starboard side. The deletion of the port side cannon allowed this space to be used to house avionics that were made homeless by the installation of a built-in flight-refuelling probe in the port upper forward fuselage.

A pair of RB199 turbofan engines powers all variants of the Tornado. The three Tornado partner nations formed Turbounion GmbH to design and build the RB199 engines to power the new combat aircraft. Turbounion is a partnership between Rolls Royce (UK) with 40%, DASA MTU (now part of EADS-Germany) 40%, and Fiat Avio in Italy with a 20% share.

The Tri-national RB199 was specifically designed to power the Panavia Tornado family of combat aircraft. For this, the engine had to be developed to meet a diversity of operational requirements, although it was primarily designed for performance at low altitudes. The compact design produces high thrust to weight and thrust to volume ratios with exceptional handling characteristics. The engine was designed to combine high power with low-fuel consumption extending aircraft range and reducing operating costs.

Turbounion succeeded in producing a compact, powerful engine optimised for low-altitude operations, although at medium to high altitudes the performance of the engine is inadequate. While the General Electric F404 class of turbofan engine is 403-cm (159-inches) in length and weighs 908-kg (2,000-lb) in weight, the RB199 is was only 324-cm (127-inches) long and weighed 900-kg (1,980-lb).

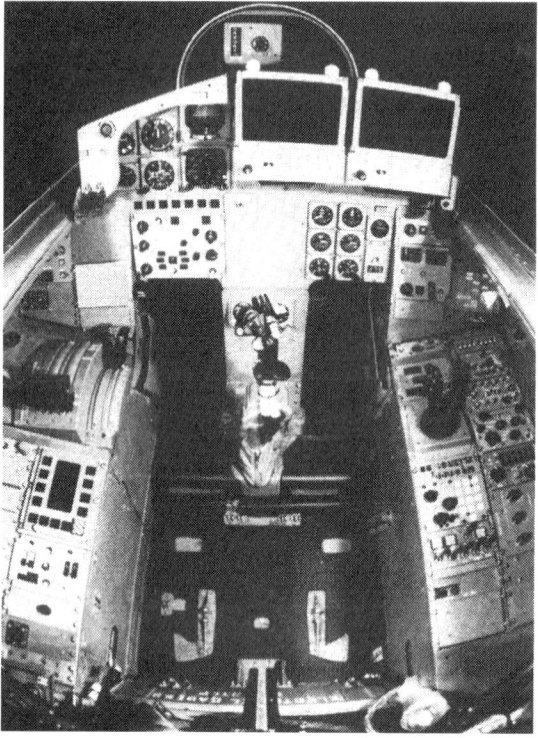

Top: The ADV cockpits would be considered positively old fashioned by modern standards. The HUD dominates the front cockpit of the Tornado F.3 and the single centre mounted monochrome TV display just ahead of the pilot's stick. Above: The rear cockpit of the Tornado F.3 is dominated by the two large format monochrome TV display screens. The screens are mounted in the centre on single-stick aircraft or offset to the right on the twin stick aircraft seen above: This allows a set of basic flight instruments to be included in the twin-stick operational trainer variant. Both BAE Systems

The RB199 is a shaft reheated turbofan engine with three low-pressure (LP), three intermediate-pressure (IP) and six high-pressure (HP) compressor stages, powered by two single-stage turbines (LP and IP) and a two-stage HP turbine. The combustion system is of annular vaporising design.

The initial production engine was the RB199 MK101, which powered early Tornado IDS aircraft. This variant gave way to the RB199 MK103, which became the standard production engine for the Tornado IDS. The first eighteen production Tornado ADV's were built to F.2 standard, powered by the RB.199 MK 103 engine. The Tornado F MK3 is powered by the higher thrust RB199 MK104. This improved variant of the engine had around 10% more thrust and was more fuel-efficient compared with the MK103. The afterburner is a compact unit and has no area where core gas and by-pass air can mix. The two mixtures are burned concurrently.

Changes to the RB199 MK104 included 36-cm (14-inch) jet pipe extension for an extra afterburner section, giving an extra 2-kN (450-lb) thrust over the MK103. The MK104 also introduced FADEC (Full Authority Digital Engine Control Unit), which was a replacement for the analogue mechanical system used in earlier variants of the RB199. FADEC is now incorporated in all production RB199's including the MK105, which was developed to power the German and Italian Tornado ECR (Electronic Combat and Reconnaissance) variant. FADEC allows better control over the engine, is more reliable and incorporates built-in test equipment.

The Tornado was one of the first combat aircraft in the world to dispense with the traditional braking parachute and instead use an integrated thrust-reverse system. In the Tornado the thrust reverse buckets, which are located just aft of the after-burner, deploy as soon as the wheels hit the ground and the spoilers come out. After deploying, the thrust reverse allows the Tornado to land in a very short distance for a front line fighter.

Turbounion RB199 Mk104
Thrust: 9,100-lbf dry and 16400-lbf with reheat.
Bypass ratio: 1.1
Pressure Ratio: 23.5
Length: 142-inches
Diameter: 28.3-inches
Basic weight: 2,151-lb
Compressor: 3LP, 3IP and 6 HP
Turbine: IHP, 1IP and 2LP

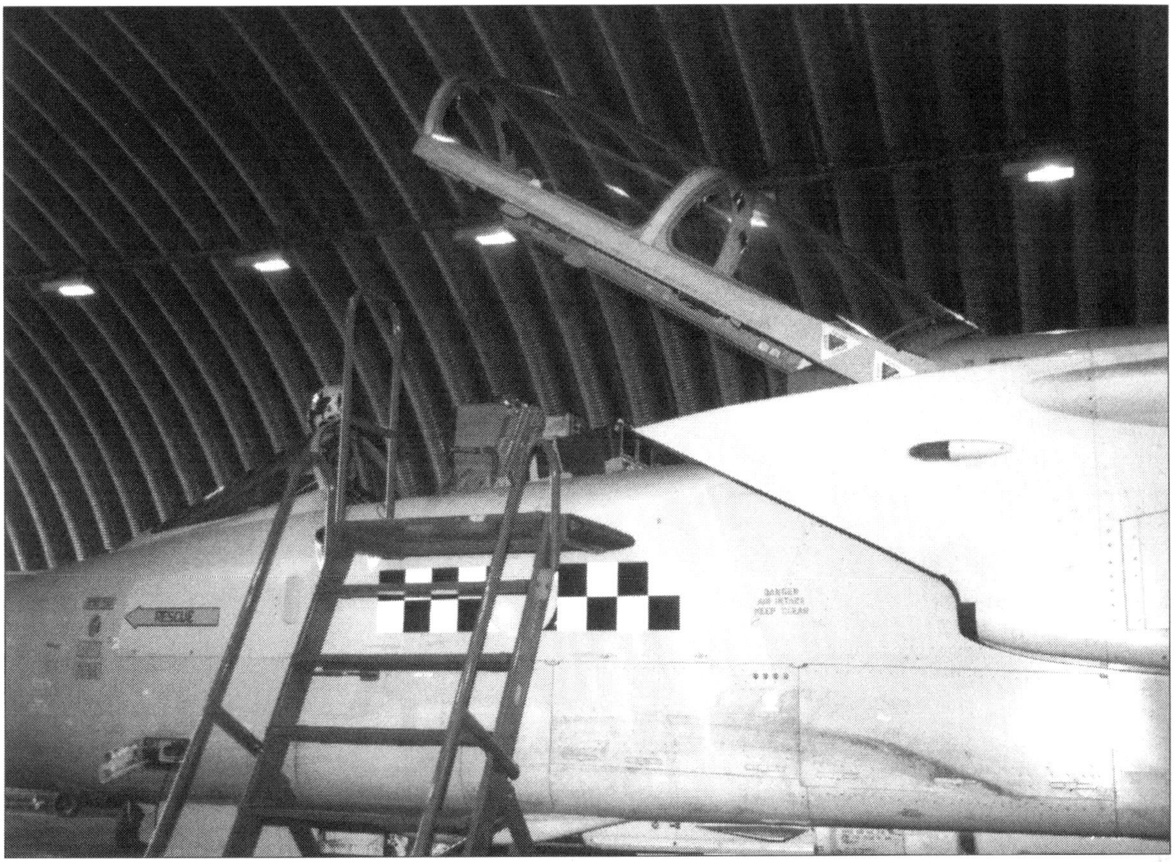
RESCUE

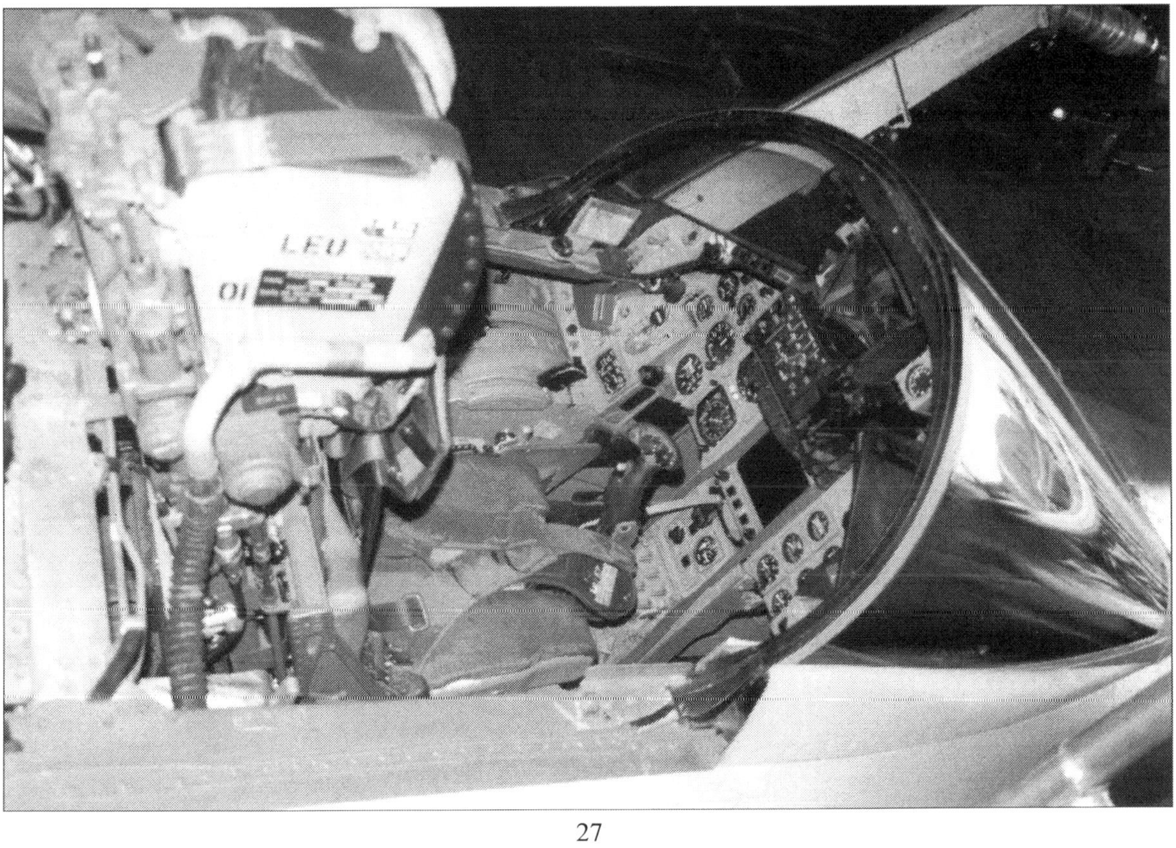
LEU
01

Above: A view of a Tornado F.3 seen from the rear cockpit of another F.3. Note the tops of the two main Monochrome TV display screens. BAE Systems **Left: The two-crew concept gives the Tornado F.3 some advantages over contemporary single-seat fighters such as the Boeing F-15C. The extra crewmember gives lower workload and increases crew situational awareness.** Crown Copyright

The Tornado ADV cockpit was designed just prior to the change to glass cockpit technology typical of modern military and civil aircraft. The dominant feature of the front cockpit is the HUD (Heads-Up-Display) which provides the pilot with all relevant flight information such as altitude, attitude and speed. The HUD also provides target engagement parameters, while a CRT (Cathode Ray Tube) HDS (Head-Down-Screen) positioned in front of the control stick displays radar and navigation information from the rear cockpit. A weapon control unit is located to the left of the HUD and a RHWR (Radar Homing Warning) Receiver display is located to the right of the HUD. Two CRT TV display screens at top centre of the dash dominate the rear cockpit for the navigator (or weapons system operator). One of the rear cockpit display screens is used mainly to display the

The combination of high power, Foxhunter radar and Skyflash BVR missiles bestowed upon the Tornado F.3 an excellent interception capability when it entered service in the 1980's. G H Lee

radar picture. The rear crewmember operates the radar during beyond visual range operations and the pilot operates combat modes during close-in engagements. The second TV display screen is used to display the Plan Display, which provides the crew with a picture of the air scene, with various data such as areas of responsibility of fighter aircraft, air to air refuelling tanker areas, coastlines and missile engagement zones. This information can be loaded by cassette on the ground before take-off or in the air. The Plan Display gave the F.3 a great advantage in fighting the air battle, with a similar system lacking on contemporary fighters such as US F-16's and F/A-18's. The plan display is now operated in parallel with the Joint Tactical Information Distribution System (JTIDS) since this system was introduced into the Tornado F.3 force. The Tornado F.2/3 twin-stick operational trainer variants have the two screens in the rear cockpit moved to the right to

allow a set of basic flying instruments to be incorporated on the dash. A throttle is located on the left side instrument panel. The JTIDS datalink along with the secure voice VHF/UHF communications suite allow the Tornado F.3 to receive voice and data communications in a secure jam-resistant environment.

In line with its different role, the Tornado ADV had the IDS Texas Instruments TNR (Tornado Nose Radar) replaced with the GEC Avionics (later GEC Marconi and now part of BAE Systems) AI.24 Foxhunter radar. The early problems have been solved with the RAF's Tornado F.3's all at least at Stage One standard by the mid-1990's. RAF Foxhunters are now at the very capable Stage 2G standard introduced from the late 1990's. The AI.24 was the first indigenous fighter radar developed in Britain for the RAF since the AI.23 Airpass radar was developed for the BAC Lightning in the 1960's. Another British fighter radar had been designed in the 1970's called Blue Fox. However, this radar was developed for the BAe Sea Harrier FRS.1 fighters for the Royal Navy and was not suitable or powerful enough for the RAF air defence role.

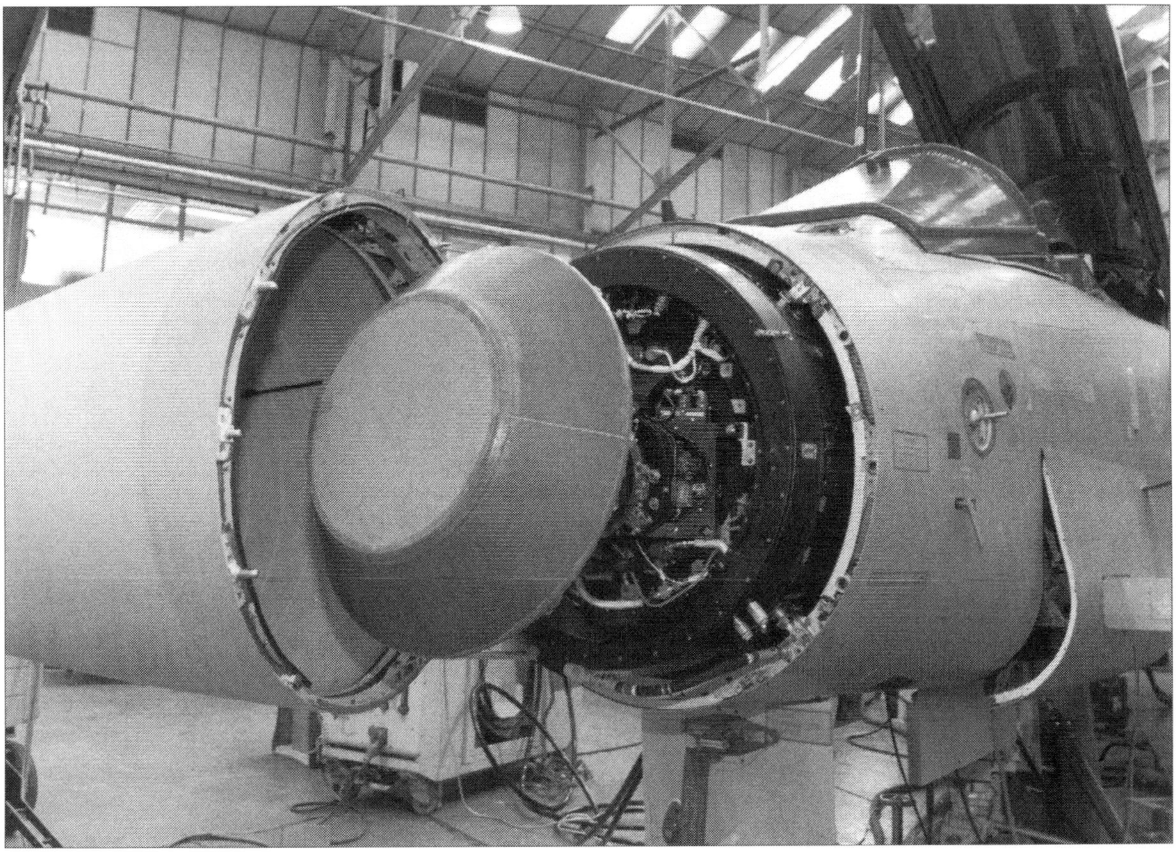

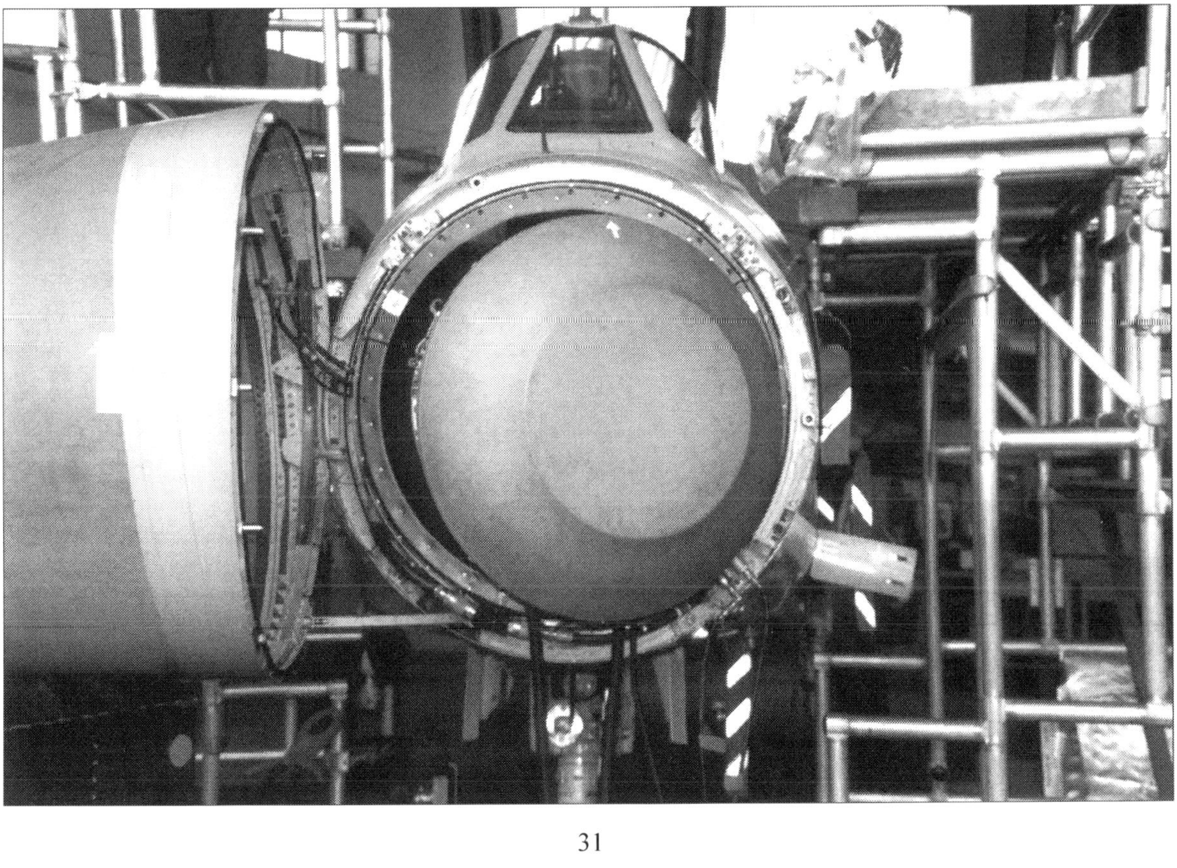

Previous page: The AI.24 Foxhunter was developed as the radar system for the Tornado ADV. Early problems associated with early Foxhunter Type W sets seen on a Tornado F.2 top were overcome by the time the RAF began to operate the improved Stage 1+ Foxhunter seen on a Tornado F.3 bottom. BAE Systems and Author

This page top: The second Tornado F.2 prototype A02, ZA267 flies alongside a BAe Lightning fighter during Foxhunter radar development flight testing from Warton. Above: The third Tornado F.2 prototype A03, ZA283 was the main radar development aircraft. Here the aircraft is taking on fuel from a RAF VC 10 tanker aircraft as three Lightning radar target aircraft sit on the tankers left side. Both BAE Systems

The Foxhunter, which benefited from development work carried out post AI.23 and pre AI-24 operates in the I-band using a pulse Doppler technique known as Frequency Modulated Interrupted Continuous Wave. The twist cassegrain antenna was designed to give good performance even in a high clutter or jamming environment. The first Foxhunters sets were flight tested in converted BAe Canberra and Buccaneer aircraft. The third prototype Tornado F.2, A03 (ZA283) was the first Tornado to be equipped with a Foxhunter radar set, flying with radar installed for the first time on 17 June 1981, following almost a year of delays. By March 1983, A03 was equipped with its third development variant of the Foxhunter known as the B series.

The Tornado F.3 was designed to meet a requirement for a long-range interceptor capable of intercepting long-range Soviet (now Russian) bombers strike aircraft and maritime patrol aircraft at ranges of around 300+ miles from the United Kingdom mainland. This pair of Tornado's from RAF Leuchars is seen escorting a Russian Tu-95 'Bear' in 1995. Crown Copyright

Top: Tornado F.3 ZE731 GF resplendent in No.43(F) squadron 80th anniversary colour scheme sits on the tarmac at RAF Leuchars in 1996. Above: With St Andrews Bay in the background, Tornado F.3 ZE858 GO from Leuchars based No.43(F) squadron taxis after a sortie at Leuchars in September 1993. Both Author

![Tornado F.3 in flight with swept wings](Tornado F.3 photograph)

Equipped with four Skyflash and Sidewinders, a Tornado F.3 from No.56(Reserve) squadron demonstrates the sleek lines of the Tornado with wings in the maximum 67 degrees swept position. BAE Systems

The AI.24 Foxhunter underwent continuous delays, resulting in the Tornado F MK2 entering service much later than was originally planned. The first of 20 pre-production Foxhunters was delivered in July 1983, although these sets were below standard and not qualified for service. It was planned for the first production F MK2s to roll off the Warton production line with Foxhunters fitted later that year. However, Tornado F.2 deliveries started from November 1984, without radar installed. Instead, the aircraft were delivered with lead ballast (known as Blue Circle radar) in their noses. The Blue Circle name indicated that the ballast was concrete, although as stated above the aircraft were fitted with lead ballast. Production Tornado F.3's did not start to come-off the Warton production line with radar fitted until the middle of 1985. After the initial contract and specification of Foxhunter had been agreed the RAF requested greater capability; this resulted in early radar sets being delivered below the specification required for operational service, contributing to the many delays encountered with the equipment.

Originally, the RAF had not required a tail chase capability for instance. Other shortcomings included large side lobes, which increased vulnerability to jamming, along with a below standard multi-mode-tracking capability. The specification had called for 20 aircraft to be tracked while the radar continued to scan. As a result of the shortcomings, GEC began a three-year period of improvements that would bring all Foxhunters up to an acceptable standard by the early 1990's.

The Tornado is in its element at lower altitudes. This F.3 is seen at low level from head-on demonstrating the sleek lines even with wings at minimum sweep back. Crown Copyright

Top: A pair of Tornado F.3's, ZE736 (HA) from No.111(F) squadron (background) and ZE256 (GA) from No.43(F) squadron (foreground) escort a Russian Air Force Sukhoi Su-27UB 'Flanker' Blue 389 into RAF Leuchars in September 1992. Crown Copyright

Above: ZE730 GL returns to Leuchars after being launched on QRA to intercept a pair of Russian surveillance aircraft in September 1993. Author

A Tornado F.3 from No.11(F) squadron manoeuvres with wings in the intermediate sweep position. The aircraft is carrying four Skyflash and two Sidewinders. G H Lee

The first production standard radar was the A1.24 Foxhunter Type W, 70 of which were built for the sixty-two Block 8-10 aircraft. A further eighty Type Z Foxhunters were delivered for eighty Block 11-12 Tornado ADV's, while the final 76 Foxhunters - 50 for 46 RAF F.3s, and 26 for 24 Royal Saudi Air Force F.3s - were delivered to Foxhunter Stage One standard. The original Type Z radar was produced to meet ASR.395, meeting the RAF's original specifications before additional capability was requested. As the Type W was below even this original standard, all but 26 sets were upgraded to Type Z in a program, which began in 1988. The Type Z sets introduced improvements including increased detection range and tracking capability. The Stage One standard of radar came closest to meeting the RAF's specification, introducing software improvements which gave much better close combat capability and improved ECCM (Electronic Counter Counter Measures). New Stage One Foxhunter sets began to be manufactured in September 1988, with 124 Type Zs Foxhunters, some of which were modified from Type W, also being brought up to Stage One standard.

The A.I24 Foxhunter uses a technique known as Frequency Modulated Interrupted Continuous Wave (FMICW), with which is integrated a Cossor IFF-3500 radar signal interrogator processor to suppress ground clutter (this was one of the problems associated with the radar's long development). Foxhunter's High Pulse Repetition Frequency (PRF) allows it to detect targets at an

The introduction of JTIDS to the Tornado F.3 force allowed aircraft to receive secure data from other sources including other fighters and the RAF's E-3D Sentry AEW MK1. Crown Copyright

initial range in excess of 100-nm while the FMICW allows the targets range to be determined from the frequency range between transmission and reception. Once a target or targets are detected, they are then stored in the central digital computer. Since the radar continues to scan normally the targets are unaware that they are the subjects of detailed analysis. The system rejects unwanted signals, leaving only real targets, which are then passed through the radar data processor before display to the aircraft crew. While the radar continues to provide data on target ranges, velocities and tracks of established targets, it continues to scan the area for additional targets.

As mentioned above, the RAF Tornado F.3's were eventually equipped with the Foxhunter Stage 2G radar, which was developed from the Stage 2AA/AB variants. The introduction of Stage 2 Foxhunter increased the F.3's detection capabilities by the addition of a new processor giving'

automatic target acquisition and tracking and discrimination between head on targets by analysis of the radar signature of their first and second stage compressor blades.

The Tornado GR.1 was equipped with an Electronica ARI 23284 Radar Warning Receiver (RWR). The Tornado F.3 was equipped with the more capable Marconi Space and Defence Systems Hermes modular Radar Homing and Warning Receiver (RHWR). The antennae for this are mounted on the upper fin trailing edge, and on the aircraft's wingtips. Threats detected by the system are displayed orally and visually on the Cathode Ray Tube Screen in both cockpits. The threats are then analysed, classified and displayed in alphanumeric form with accurate range and bearing of the threats. This is then compared with the Foxhunter radar target tracks, therefore, allowing the crew to classify and allocate priority to any potential threats. The system can act passively or can burn through enemy jamming.

The Tornado F.3 is now equipped with the JTIDS (Joint Tactical Information Distribution System), which provide the Tornado F.3 crew with an up to date picture on the entire air battle. JTIDS allows

JTIDS enables the Tornado to relay fuel and weapon states to other platforms as well as receive information on other air and surface based platforms. A pair of Tornado F.3's take-on fuels from a RAF VC 10 tanker aircraft. The aircraft are armed with four Skyflash and four ASRAAM air to air missiles and carrying a Phimat countermeasures dispenser pod on the right outboard wing station and an Ariel towed decoy pod on the left outboard wing station. Crown Copyright

the F.3 to contribute its own information, such as fuel and weapon states back to their tactical headquarters. JTIDS provides the F.3's and other friendly forces with a secure jam resistant communication network that can operate in a dense ECM (Electronic Counter Measures) environment. The system which was manufactured by Singer Kearfott and Rockwell Collins in the US and GEC Marconi in the UK, allows secure voice communications, accurate navigation and identification of friend or foe.

The Joint Tactical Information Distribution System is a Time Division Multiple Access (TDMA) communication system, which operates at L-band frequencies. The system operates over line-of-sight at ranges up to 500 nautical miles with automatic relay extension beyond. JTIDS allows various assets to contribute and receive data via

jam resistant voice and data communication. Information can be used for command and control, navigation, relative positioning and identification. As well as distribution of tactical information in digital form, JTIDS also locates and identifies subscribers with respect to other users. The system is highly resistant to jamming due to spread-spectrum and frequency hopping techniques and data encryption allows the secure transfer of information. The system can handle large amounts of data and voice communications and can automatically broadcasts outgoing messages at pre designated and repeated intervals. When not transmitting, the JTIDS terminal receives messages from other JTIDS terminals, which also transmit in prearranged order.

The second production Tornado F.3 ZE155 was held back on the Warton production line to be equipped with JTIDS. The aircraft flew on 16 October 1986 with a JTIDS Class 2 terminal installed. Initial trials of the system were conducted in January 1987, when the aircraft established secure voice-free text and fixed format communications with a ground based JTIDS terminal. In September 1987, ZE155 went to Yuma, Arizona in the USA to conduct further JTIDS trials becoming the first Tornado F.3 to cross the Atlantic with in-flight refuelling provided

A No.11(F) squadron Tornado EF.3 goes vertical for the camera armed with four ASRAAM air to air missiles on the inboard wing stub stations and a pair of ALARM 2 anti-radiation missiles on the under fuselage stations. The aircraft is also carrying a pair of 1500-litre external fuel tanks on the inboard stations; a Phimat chaff dispenser on the right outboard station and an Ariel towed radar decoy on the left outboard wing station. G H Lee

by a RAF Lockheed Tristar K MK1 tanker aircraft. On 24 September 1987, ZE155 became the first British fighter aircraft to cross the Atlantic without in-flight refuelling. The aircraft flew direct from Goose Bay in Canada to Warton in the UK, a distance of 2,200 miles in a time of 4-hours and 45-minutes.

Introduction of JTIDS to the operational Tornado F.3 fleet commenced in 1994, with aircraft with serials from ZG751 to ZH558. All initial JTIDS equipped Tornado's were initially based at Coningsby before allocation to other stations. In August 1994, a quartet of Tornado F.3's from No.5 squadron and the Tornado F.3 OEU at Coningsby, together with a RAF Boeing E-3D Sentry AEW MK1 from No.8 squadron RAF Waddington

deployed to Mountain Home AFB, Idaho, USA with in-flight refuelling support from RAF Tristar K MK1 tankers. The three week deployment was used to test the inter-operability of JTIDS between differing aircraft types including the only USAF fighter unit then equipped with JTIDS, the 390th FS (Fighter Squadron) equipped with Boeing F-15C Eagles.

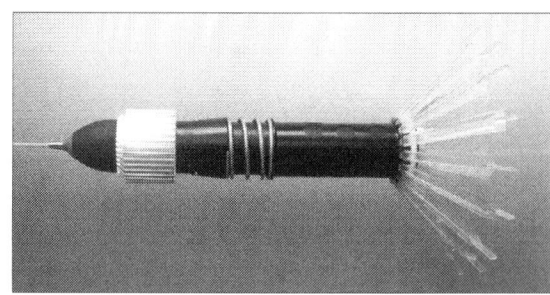

The Ariel towed radar decoy above is designed to protect the mother aircraft from radar guided surface to air and air to air missiles. The system was introduced to Tornado F.3's from the mid-1990's. BAE Systems

Tornado F3's took on the SEAD role armed with MBDA ALARM 2 (Air Launched Anti-Radiation Missiles). The aircraft are sometimes referred to as EF.3. This aircraft is carrying two ALARMs, four ASRAAM's, external fuel tanks, Phimat pod and Ariel towed decoy. G H Lee

The Ariel towed radar decoy was introduced to service on RAF Tornado F.3s operating over Bosnia Herzegovina in 1995. The decoy uses a techniques generator as a decoy to counter missile threats. The variant of decoy deployed on RAF F.3s was thought to have been optimised for countering semi active radar guided missiles such as the AA10 'Alamo' as deployed on Serbian Air Force MiG-29 'Fulcrums'. The requirement for a decoy probably originated from the requirement to counter passive AA10 missiles, which home in on the target aircraft radar emissions. The decoy would then simulate the parent aircraft fooling the missile.

A MBDA Skyflash beyond visual range air to air missile is launched from a Tornado F3 during weapons trials over the Aberporth ranges. BAE Systems

When it entered service, the Tornado F3's primary armament consisted of four BAe Dynamics (now Matra BAe Dynamics Alenia) Skyflash SARH (Semi-Active Radar Homing) air-to-air missiles under the fuselage. Skyflash had replaced the earlier US Hughes (now Raytheon) AIM7E-2 Sparrow on the Phantoms, before being carried over to the Tornado F2/3. The Skyflash was based on the body of the Sparrow, but with an entirely new British motor and seeker head. The Skyflash was exported to Sweden, which used the weapon on the Saab JA 37 Viggen fighters in service with the Flygvapnet (Swedish Air Force), under the designation RB 71.

Skyflash was designed as a supersonic medium range air to air missile. The latest variants of the missile incorporated a boost-sustain, solid fuel rocket motor increasing range over earlier variants. The missile can be employed against targets in all-

weather conditions and has the ability to snap-up or snap-down allowing engagement of targets at 'ultra-high' or 'ultra-low' levels. The seeker-head can discriminate between separate target groups and can operate in a dense counter measure environment.

Flame and smoke erupts from the rocket motor exhaust of a Skyflash air to air missile as it is launched from a Tornado F3 from No.111(F) squadron. Crown Copyright

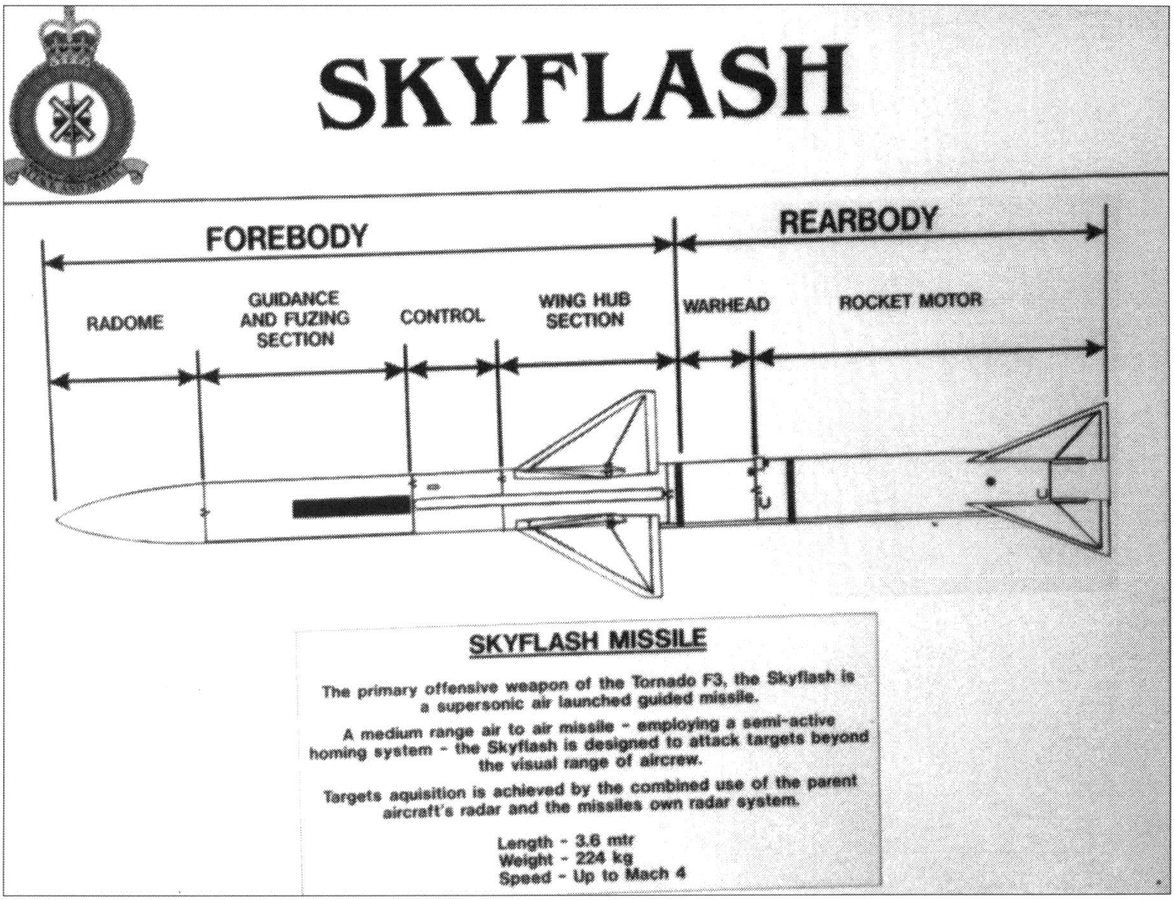

SKYFLASH

FOREBODY   REARBODY

RADOME   GUIDANCE AND FUZING SECTION   CONTROL   WING HUB SECTION   WARHEAD   ROCKET MOTOR

SKYFLASH MISSILE

The primary offensive weapon of the Tornado F3, the Skyflash is a supersonic air launched guided missile.

A medium range air to air missile – employing a semi-active homing system – the Skyflash is designed to attack targets beyond the visual range of aircrew.

Targets aquisition is achieved by the combined use of the parent aircraft's radar and the missiles own radar system.

Length - 3.6 mtr
Weight - 224 kg
Speed - Up to Mach 4

This schematic of a MBDA Skyflash BVR air to air missiles shows the two main sections, which are broken down into six sub-sections. The specification information from RAF supplied information shows a speed on up to Mach 4, although other information literature supplied by the MoD shows lower speeds.
Author

The Tornado F2/3 can carry four Skyflash missiles on the under-fuselage stations, with the front pair of missiles being semi-recessed. When launched, the missiles are pushed downward into the airflow on large Frazer Nash rams. Once launched, the missiles uses semi-active radar homing, whereby it heads towards the target area guided by the Tornado's AI.24 Foxhunter radar. The launch aircraft has to continually illuminate the target allowing the Skyflash to home in on the targets radar returns. Once near the target, Skyflash detonates using a proximity fuse. Although primarily used as a BVR weapon, Skyflash can also be used against targets at shorter ranges. In this type of engagement, the missile is optimised for quick reaction and maximum manoeuvrability once launched from the mother aircraft.

In February 1987, the Tornado F.3 conducted missile firings at the Aberporth ranges for the first time, achieving good results in Skyflash firings against target drones. The missile has been deployed operationally with F3's from the RAF, RSAF and Italian Air Force in the Persian Gulf and the Balkans, but none was ever fired in anger.

This No.11(F) squadron Tornado F3 is armed with two Skyflash missiles in forward fuselage recesses.
Crown Copyright

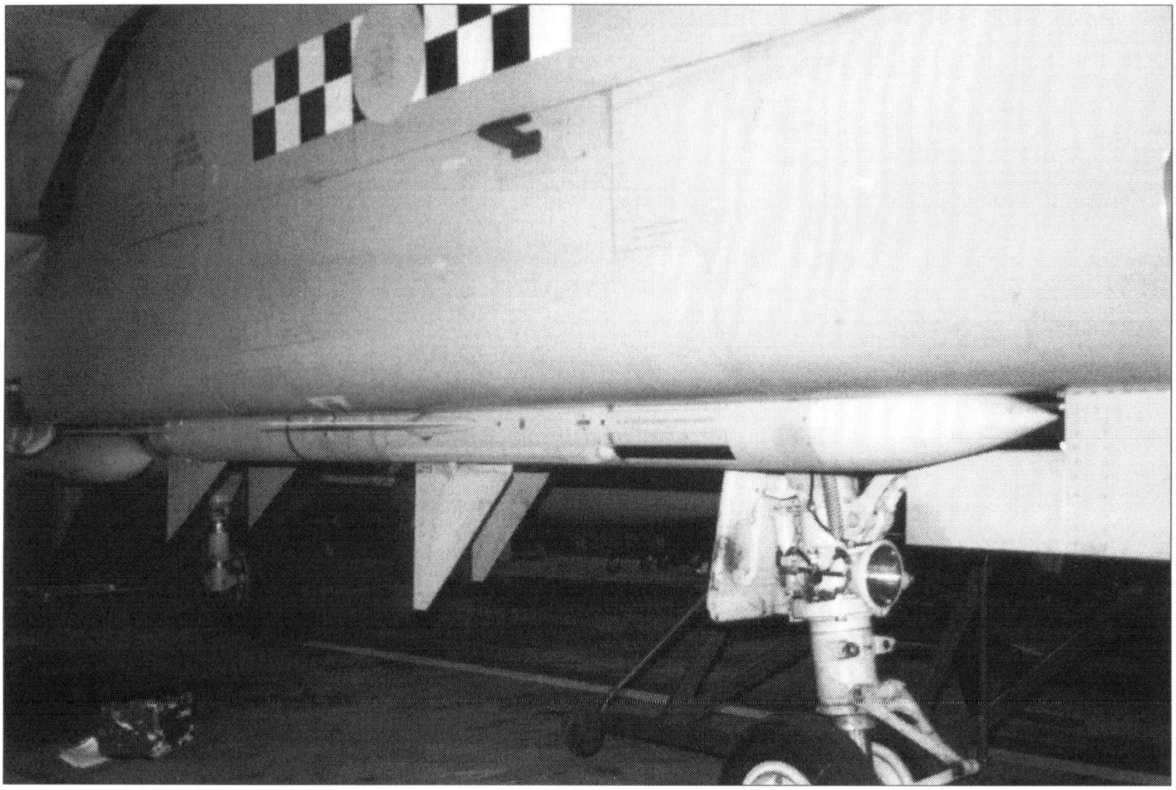

Top: Skyflash missiles in forward recesses of a Tornado F3 on QRA at Leuchars. Above: Skyflash **and Sidewinder missiles are displayed alongside a No.111(F) squadron F3 in 2002.** Both Author

This Tornado F.3 is armed with a full complement of four Skyflash, but only two AIM-9L Sidewinders on the outer stub pylons. G H Lee

Initial plans to develop the missile into a fire and forget weapon under the Active Skyflash program were eventually abandoned when the RAF began to lean toward the US Hughes (now Raytheon AIM-120 Advanced Medium Range Air to Air Missile (AMRAAM).

MBDA Skyflash Semi-Active Radar Homing Medium Range Air to Air Missile

Length: 3.66-metre
Diameter: 20.3-cm
Launch weight: 208-KG (RAF information also quotes a weight of 224-kg).
Range: Around 30+ miles
Speed: Up to Mach 4
Guidance: Uses launch aircraft radar for initial guidance before switching to on-board active radar terminal guidance radar seeker

When it entered service, the Tornado F.2/3 was armed with four Raytheon AIM-9L Sidewinder air-to-air missiles as its primary short-range infrared guided air to air missile system. The Sidewinder, like the Skyflash, was carried over from the RAF Phantom fleet. The AIM-9, which is an infrared (heat-seeking) guided missile designed for within visual range air to air combat, has undergone continued development since the first models entered service in the 1950's. The missile is made-up of a rocket motor, a high explosive warhead and an active optical tracking system. The infrared guidance gives the missile its fire and forget capability, allowing the launch aircraft to leave the area or take evasive action after launch.

The RAF uses the AIM-9L and AIM-9M variants of the missile, which offer improved resistance to infrared countermeasures over earlier Sidewinder variants. After launch, the missile uses its infrared seeker-head to home in on the target. Once near the target, the proximity fuse detonates while the target is inside the lethal area. The benefits of infrared guided missiles in close-range air to air combat include fire-and forget capability and quick-reaction time.

The AIM-9L was the standard variant of the AIM-9 Sidewinder to equip Tornado F.3 squadrons when the type entered service in the 1980's. From 1990, some Tornado F.3's were armed with the improved AIM-9M variant of the missile. Tornado F.3 ZE200 DB in No.11(F) squadron markings launches an AIM-9L Sidewinder, which is just leaving the launch rail top and streaks away, engulfing the launch aircraft in smoke above. Both BAE Systems

An AIM-9M Sidewinder on the LEU 449 launch rail fitted to the inboard wing station stub pylon of a No.43(F) squadron Tornado F.3 in the QRA shelter at RAF Leuchars in 1993. Author

While the Tornado IDS was equipped with a pair of 27-mm Mauser cannon - one on each side of the forward fuselage - the Tornado ADV is armed with only a single cannon housed in the starboard lower fuselage.

One of the major drawbacks of semi-active radar guided air to air missiles is the requirement for the launch aircraft to continually illuminate the target with its radar for the entire flight of the missile. This leaves the launch aircraft vulnerable to enemy counters including air launched BVR and short range missiles as the distance between the target and the launch aircraft closes at high speeds sometimes bringing the launch aircraft within range of short-range infrared guided air to air missiles. To overcome this vulnerability, the RAF had long harboured plans for introduction an active beyond visual range air to air missile to replace the Skyflash on the Tornado F.3. This was initially planned to be an active radar guided variant of the

Skyflash itself, most commonly referred to as 'Active Skyflash'. However, this program was never brought to fruition and RAF interest switched to the US Raytheon AIM-120 AMRAAM active radar guided beyond visual range air to air missile, which was entering service on US fighters in the early 1990's.

A Pair of Skyflash and Sidewinder armed Tornado F.3's fly in formation with a RAF Hawk T MK1 advanced jet trainer. Crown Copyright

From mid-2004, the Raytheon AIM-120 AMRAAM has been fully operational with No.43(F) and No.111(F) squadrons based at RAF Leuchars. This pair of Tornado F.3's from No.43(F) (foreground) and No.111(F) squadrons (background) seen in summer 2004 is armed with four AMRAAM on the fuselage stations and four ASRAAM's on the inboard wing station stub pylons. Crown Copyright

Plans to integrate AMRAAM onto the RAF's Tornado F.3 force were formulated in the early 1990's, but were dropped in 1992 as the RAF looked toward replacement of the Tornado F.3 with the European Fighter Aircraft (Later Eurofighter 2000 and now Eurofighter Typhoon). However, with delays in the planned introduction to service of the Eurofighter slipping beyond the year 2000, plans to integrate AMRAAM along with the MBDA ASRAAM were resurrected as part of the Tornado F.3 Capability Sustainment Program (CSP). This was initiated in 1996, to keep the Tornado F.3 a viable air defence platform until final retirement planned for 2010.

Designed as a direct replacement for the AIM-7 Sparrow semi-active radar homing missile, the AMRAAM has revolutionised NATO air to air capability. Unlike the vintage Sparrow, which required the launch aircraft to continually keep the target illuminated by its radar during the missile flight time, the AMRAAM has its own active-homing radar on board, together with an INS and a datalink. Once launched, AMRAAM will fly towards the target controlled by the pre-programmed inertial guidance system, requiring no further assistance from the launch aircraft. For longer-range engagements the operator can provide mid-course guidance, transmitting target location data to the missiles guidance system, following which, the missile goes into the autonomous mode, using inertial guidance only. When it nears the target area it switches to the terminal homing mode, using the missiles own active monopulse radar seeker. AMRAAM can be used to engage multiple targets simultaneously. With the Sparrow or British Skyflash, only one target could be engaged at a time.

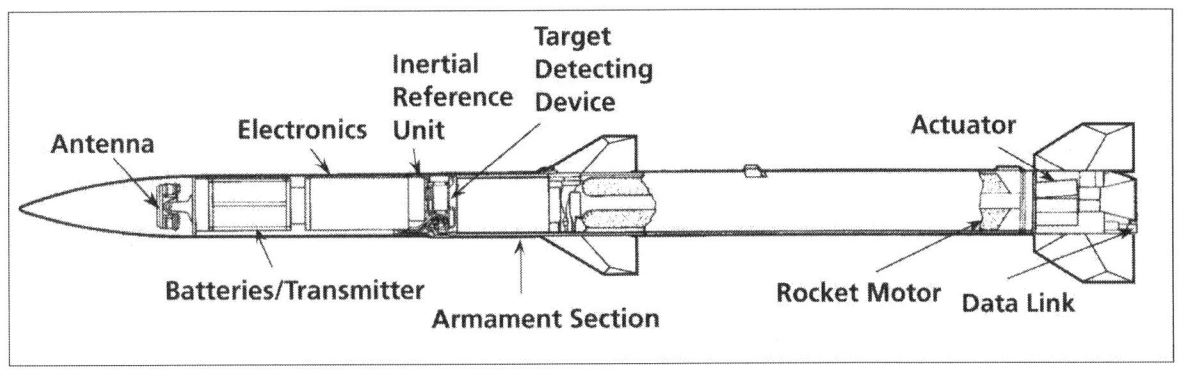

Top: Cutaway drawing of the AIM-120 AMRAAM showing the main components. Raytheon **Above: A Tornado F.3 from the F.3 OEU (Operational Evaluation Unit) (foreground) conducts flights carrying AMRAAM and ASRAAM missiles. The centre aircraft in No.5 squadron markings is armed with Skyflash and Sidewinders, while the farthest aircraft appears to be unarmed.** Crown Copyright

The AMRAAM has a similar configuration to Sparrow with four fixed central wings and four rear control fins. AMRAAM is faster with a longer engagement range, is more manoeuvrable than its predecessor and is also more resistant to ECM. As long as the launch aircraft has a track while scan radar up to six missiles can be fired in rapid succession against separate targets. The missile is smaller and lighter than its predecessor with a length of 3.65-m, a diameter of 17.8-cm, a span of 63-cm, and weighs in at 157-kg (345.4-lb), including the 23-kg (50.6-lb) pre-fragmented High Explosive warhead with either a proximity or contact fuse. Powered by a solid propellant rocket motor, AMRAAM has a speed of Mach 3 and a maximum range of 65-km (40.4 miles). AMRAAM can operate in three modes, depending on target range and conditions of engagement.

Once launched, it is designed to fly out towards the target under the control of the pre-programmed inertial guidance system needing no further assistance from the launch aircraft. For long-range engagements, the aircraft can update the missile flight path mid-course, transmitting target location data to the missile guidance package. After this the missile goes into the autonomous mode, using inertial guidance only. When it nears the target area it switches to the terminal homing mode using its own on board active monopulse radar seeker. Under its original guise, the Tornado F.3 CSP upgrade did not provide for mid-course guidance, however, this short-sighted error has been addressed under the AMRAAM optimisation program. The Pre-Planned Product Improvement (P3I) program is designed to ensure that capability is sustained throughout its service life. RAF AMRAAM rounds are slightly downgraded compared with rounds in US service.

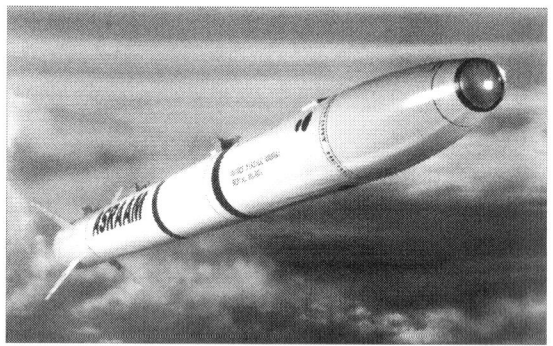

An AMRAAM drill round sits in front of Sidewinder and Skyflash drill rounds in a hanger at RAF Leuchars in September 2002. Author

AMRAAM entered service in time to be employed operationally during the 1991 Gulf War; however, no missiles were launched before the cease-fire. The first combat use of AMRAAM occurred on 27 December 1992, when a USAF Lockheed Martin (then General Dynamics) F-16D Block 42 fighter launched an AMRAAM from BVR, destroying an Iraqi MiG-25 'Foxbat' E in the controversial NFZ (No Fly Zone) over Southern Iraq. This NFZ was initially set-up by France, the UK and US, but not sanctioned by the United Nations. AMRAAM has been used in several engagements by US and NATO forces since 1992, specifically against Iraqi and Serbian aircraft.

When introduced to service, AMRAAM was rightfully regarded as a quantum leap in capability over the vintage AIM-7 Sparrow that it was replacing. While the missile has been successfully used during a number of operations and its capabilities demonstrated in numerous test firings, there has long been concern over an apparent drop-off in speed during long range flights following rocket motor burn out. As the speed falls off, the missile becomes less capable of intercepting a manoeuvring target. After launch, during rocket motor burn, AMRAAM accelerates to Mach 3, but following rocket motor burn out speed gradually decreases to between Mach 1 to Mach 1.5. The deficiencies were highlighted in January 1999 when a number of missiles were fired at Iraqi MiG's during two separate engagements, all of which failed to achieve a kill. It is thought that as speed fell off the Iraqi MiG's simply out ran the missiles.

Procured as a direct replacement for the AIM-9L/M Sidewinder in RAF fighter squadron service, the MBDA ASRAAM entered service on RAF Tornado F.3's in 2002. MBDA

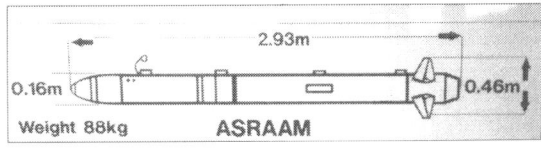

An ASRAAM training round sits along side a CSP modified Tornado F.3 from No.111(F) squadron at Leuchars in 2002. Author

From summer 2004, AMRAAM was fully operational on RAF Tornado F.3's modified under the CSP. Although CSP modified aircraft and AMRAAM stocks had been available for a number of years prior, RAF Tornado F.3's deployed on operations and conducted QRA alerts armed with Skyflash and from 2002, ASRAAM missiles. The RAF deployment to the Persian Gulf for the controversial March 2003 invasion of Iraq was armed with ASRAAM and Skyflash. The omission of AMRAAM led to some reports, even as late as early 2004 to conclude that Skyflash were superior to AMRAAM in its basic configuration. However, this totally ignored the requirement for new tactics, training and operational firings of the missile to be conducted before the weapon could be cleared for operational use. Some official documents cite a higher top speed for Skyflash of up to Mach 4, while AMRAAM has a top speed of around Mach 3. However, even in its basic configuration, the fire and forget capability of AMRAAM enhances the launch aircraft's ability to conduct counters to enemy counter fire. As with Skyflash before it, the

RAF tactical doctrine includes the launch of ASRAAM missiles at the target aircraft as it conducts a missile manoeuvre break to avoid the initial AMRAAM launch.

Diagram of MBDA ASRAAM showing basic dimensions and weight. Author

For shorter range air to air combat the Matra BAe Dynamics Alenia (MBDA) Advanced Short Range Air to Air Missile (ASRAAM) is now employed by RAF Tornado F.3's and will arm Eurofighter Typhoons T.1/A F.1's and BAE Systems Harrier GR.7/9 in RAF service. ASRAAM is replacing the AIM-9L/M Sidewinder in RAF fighter squadron service, satisfying SR(A) 1234.

On 20 September 1990, The MoD Equipment Policy committee gave the go-ahead for the ASRAAM, allowing prime-contractor BAe Dynamics (now MBDA) to launch the development program for the missile. It was

This Tornado F.3 is heavily configured for the SEAD (Suppression of Enemy Air Defences) and air-to-air role armed with MBDA ASRAAM and ALARM 2 missiles. The aircraft is also configured with a pair of 1500-litre external fuel tanks on the inboard wing stations and Phimat chaff dispenser and Ariel towed radar decoy on the outboard stations. G H Lee

planned for the ASRAAM to be the NATO Sidewinder replacement and to be procured by the US services as the AIM-132. Following the German, Norwegian and Canadian withdrawal from the program and the US decision to develop the Sidewinder further, in November 1990, the MoD invited bids from contractors to supply a Sidewinder replacement. Bids were to be submitted by 6 August 1992.

As well as BAe Dynamics/Hughes (now Matra BAe Dynamics Alenia/Raytheon) with the ASRAAM, the GEC Marconi/MATRA MICA Short Range Air to Air Missile (SRAAM) was again put forward as were designs from Loral and Raytheon in the US and Germany's BGT. At least two of these proposals were based on upgrades of the Sidewinder. On 3 March 1992, the British defence minister announced that the ASRAAM had won the competition and BAe Dynamics had been awarded a contract for production of the missile.

At the cutting edge of SRAAM (short-range Air to Air Missile) technology, ASRAAM is an extremely agile wingless, low drag missile and uses a combination of body lift and rear aerodynamic control to achieve its high performance and agility. Controlled by a software-driven autopilot, ASRAAM maintains its high agility throughout the missile flight, unlike Thrust Vector Controlled missiles. Propulsion is provided by a Royal Ordnance smokeless (low signature) solid fuelled rocket motor, which accelerates the missile to very high speed, claimed as being "hypersonic" and the fastest speed in its category.

For control, the missile employs four small control fins at the rear of the body and the missile produces less drag than other SRAAMs currently under development such as the IRIS-T, which is under development as a replacement for the Sidewinder in German Luftwaffe service. The low drag coefficient gives ASRAAM a longer engagement range than other contemporary systems. While missiles in the Rafale Python 4 class offer near instantaneous manoeuvrability off the launch rail, these weapons are very short legged compared to ASRAAM and the longer motor burn of ASRAAM

Top: An ASRAAM drill round used for training missile procedures at RAF Leuchars in 2002. Above: An ASRAAM round fitted on an inboard wing station stub-pylon on an F.3. Author and Crown Copyright

The MBDA ALARM missile was previously the preserve of the Panavia Tornado GR.1/4, which normally carried three missiles on fuselage missile stations. Tornado GR.1's can also carry ALARM missiles on the inboard wing stub pylon Author

reduces its turning circle for close-range high off-boresight engagements. A Thorn EMI active IR, laser proximity and impact fuse actuates the DASA (now EADS Germany) developed warhead. This increases the missile effectiveness against small targets such as cruise missiles.

In modern within visual range air combat, the ability to strike first is vital. ASRAAM provides an advanced capability to defeat any potential adversary, including high speed, reaction time and agility to ensure the maximum kill potential, with survivability. From close-in combat ranges to well beyond visual range, ASRAAM provides all-round target designation with full acquisition anywhere in the forward hemisphere, and the option of lock-on-after-launch engagement in both forward and rear hemispheres.

Long range and short range target acquisition and track are achieved by its Raytheon developed wide off-boresight gimballed 128 x 128-element focal

plane array imaging infrared seeker and state of the art image processor. The seeker provides real-world imagery, extended acquisition range and unprecedented countermeasure resistance.

For maximum operational flexibility, ASRAAM can be cued to the target using the launch aircraft's radar, a Helmet Mounted Sight (HMS) or by Data Link at off-boresight angles of up to 90 degrees. However, the Tornado F.3 CSP aircraft were not equipped with a HMS system; therefore, ASRAAM's high off-boresight potential could not be fully exploited by the Tornado launch aircraft.

ASRAAM can also operate in Scan Mode, providing the pilot with an autonomous Pseudo (IRST) Infrared Search and Track system. This facility increases the Tornado F.3's capability as the aircraft previously lacked an IRST capability. The missiles high speed, combined with unique all round targeting capability can permit destruction of the enemy target before they can launch their missiles. The lock-on after launch capability allows the missile to engage targets approaching from behind and the missile can regain target lock-on from the last known track even if the targets IR signature has been lost.

ASRAAM was initially scheduled to enter service with RAF Tornado F.3 interceptor squadrons from 1998. However, the program suffered a number of delays and in summer 1999 the prime-contractor offered the UK MoD an improved processor, which would enhance the missiles capability. The seeker head would be given a better field of view and the missile's maneuverability would be enhanced.

By early 2002, the first ASRAAM rounds had been received by the RAF and Tornado F.3 squadrons began training with the new missile in April 2002. RAF Harrier GR.7/9 squadrons will also be equipped with ASRAAM and the missile has been ordered by Australia for service on upgraded RAAF (Royal Australian Air Force) Boeing F/A-18A/B Hornet strike fighters.

The first RAF Typhoons delivered in 2002 and 2003 have already begun flying with ASRAAM training rounds as part of the build up to operational capability of the Tranche 1 Typhoon.

The ASRAAM development offered unsuccessfully to meet the US AIM-9X requirement featured thrust vectoring to increase its manoeuvrability in high off-boresight engagements. This could feature in possible future variants of ASRAAM.

ASRAAM

Weight: 88-kg
Length: 2.90-m
Diameter: 0.166-m
Warhead: fragmentation explosive
Fuses: laser proximity and impact
Propulsion: solid rocket motor
Homing head: Imaging infrared 128 x 128-element focal plane array

---

From early 2003, some RAF Tornado F.3's operating with No.11(F) squadron at RAF Leeming have been tasked with the Suppression Of Enemy Air Defence (SEAD) role armed with the MBDA Air Launched Anti Radiation Missile (ALARM 2). The ALARM missile, unlike the US AGM-88 HARM does not rely on onboard aircraft computers and is, therefore, self-contained. The missile can be used to attack fixed site and mobile SAM's sites identified and located by the launch aircraft. A feature unique to ALARM is the so called 'stealth' mode when the missile once launched zooms to around 70,000-ft altitude flying over the suspected Surface to air Missile (SAM) site before deploying its onboard parachute and

slowly descending waiting for an enemy radar to be switched on. This capability was designed into the missile in order to defeat the age old trick of the SAM operator switching off their radar once the SEAD aircraft has been detected only to switch it on when the coast is clear. Anti-Radiation Missiles (ARM) home in on the electromagnetic signals emitted by radar. A favoured anti ARM tactics is for the SAM crew to simply switch off their radar after which most ARM break lock. This tactic has been used countless times with enormous success to defeat the US Raytheon AGM-88 HARM (High Speed Anti-Radiation Missile). The small parachute deployed by ALARM allows the missile to slowly descend in case the radar is switched back on, after which the parachute is jettisoned and the missile once again homes-in on the radar emissions.

ALARM has an internal memory, which will guide the missile toward the target after it has stopped transmitting although this mode of attack can be severely inaccurate as opposed to the direct homing mode.

Once the ALARM role was relinquished by Tornado F.3, it still remained firmly in the hands of the Tornado GR.4. This GR.4 is armed with MBDA Brimstone air to surface missiles and ALARM 2 anti-radiation missiles. MBDA

Chapter Three

Tornado ADV – RAF Service

Above: Tornado F.3 (ZE760) from the Tornado Operational Conversion Unit (No.65Reserve) squadron is seen in 1991 armed with a load of four Skyflash and 4 AIM-9L Sidewinder missiles. BAE Systems

The first production standard Tornado F.2 to fly was actually the second production aircraft ZD900. This aircraft conducted its maiden flight from Warton on 5 March 1984. The first production F.2, (ZD899) conducted its first flight a few weeks later on 12 April 1984.

The first Tornado F.2 to enter RAF service was ZD901, the third production Tornado F.2, which joined No.229 Operational Conversion Unit (OCU) on 5 November 1984. The F.2 was always considered an interim model and only 18 were built, with the last being delivered to the OCU in October 1985. The first standard production Tornado F.3 to fly was the second production F.3, (ZE155), which performed its maiden flight on 16 October 1985. The first production standard Tornado F.3, (ZE154) conducted its first flight on

20 November 1985. This aircraft entered service with No.229 OCU in 1986, before being allocated to No.29 squadron. The RAF took delivery of its last Tornado F.3, ZH559 in March 1993 bringing to an end production of 18 F.2's and 155 F.3's for the RAF. ZH559 was one of eight Tornados F.3's originally ordered for the Royal Air Force of Oman, but delivered to the RAF.

Unofficially formed at RAF Coningsby on 1 November 1984, No.229 OCU received its first Tornado F.2's on 5 November when ZD901 and ZD903 landed at the station. Official formation date was 1 May 1985 and the unit was immediately in the limelight when one of its aircraft was put on the air show circuit, flying a synchronised display with a Supermarine Spitfire of the Battle of Britain Memorial Flight. No.229 OCU received its full complement of 16 Tornado F.2's when the last arrived on 21 October 1985. Also on this day, the Tornado F.2 was involved in its first Air Defence exercise, Priory 85/2.

The first production Tornado F.2 was ZD899 seen during a test flight in 1984. BAE Systems

Delivery of Tornado F MK.3's began with the arrival of ZE159 on 28 July 1986. With the introduction to service of the improved F.3 the F.2's were gradually withdrawn and put into storage at RAF St Athan. No.229 OCU was disbanded in July 1992 and was replaced by the Tornado F.3 OCU, which adopted the No.56(R) squadron number.

Top left: One of the first Tornado F Mk.2's to arrive at Coningsby for service with No.229 OCU was ZD903, seen arriving at the station on 5 November 1984. RAF Above: The first production Tornado F Mk.3 was ZE154, seen during an early test flight. BAE Systems

The first operational unit to equip with the Tornado F.3 was No.29 squadron, which stood down with its McDonnell Douglas Phantom FGR.2 fighters on 1 December 1986 and was declared operational on the Tornado F.3 on 1 November 1987. No.29 squadron had since 1 January 1980 been assigned to Supreme Allied Commander Atlantic (SACLANT) for maritime defence of the fleet. The squadron also had an out of NATO area commitment and was chosen for operation Golden Eagle, a circumnavigation of the world between 21 August and 26 October 1988. During this, four aircraft participated in a Malaysian air defence exercise as well as visiting some South East Asian air forces and flying at air shows in Australia as part of that country's Bi-centenary celebrations.

No.5 squadron was the second operational squadron to form on the Tornado F.3 and the first to convert from BAe Lightning F.6 fighters. The squadron re-formed on 1 January 1988 at Coningsby, although its first aircraft, ZE292 arrived at Conignsby on 25 September 1987.

The next squadron to receive Tornado F.3's was No.11 squadron, which had disbanded at Binbrook with Lightning fighters in March 1988. The first Tornado F.3 for No.11 squadron was ZE764, which arrived at Coningsby on 25 April 1988. Here No.11D squadron came into existence on 1 May 1988. On 1 July 1988, the squadron officially re formed on the F.3 after relinquishing its BAe Lightning fighters at RAF Binbrook. At the same time, the squadron moved to its new home of RAF Leeming to begin formation of the Leeming wing.

No.11 squadron commenced their work-up flights on the Tornado F.3 on 1 July 1988 and was declared operational on the Tornado F.3 and assigned to SACLANT on 1 November 1988. That same day, No.23 squadron, the second of the Leeming wings Tornado F.3 units was formed with the first F.3's for the squadron arriving at Leeming on 5 August 1988. The squadron had disbanded with its Phantom FGR.2 fighters in 1998 after No.23 squadron based in the Falkland Islands was reduced to flight of four aircraft.

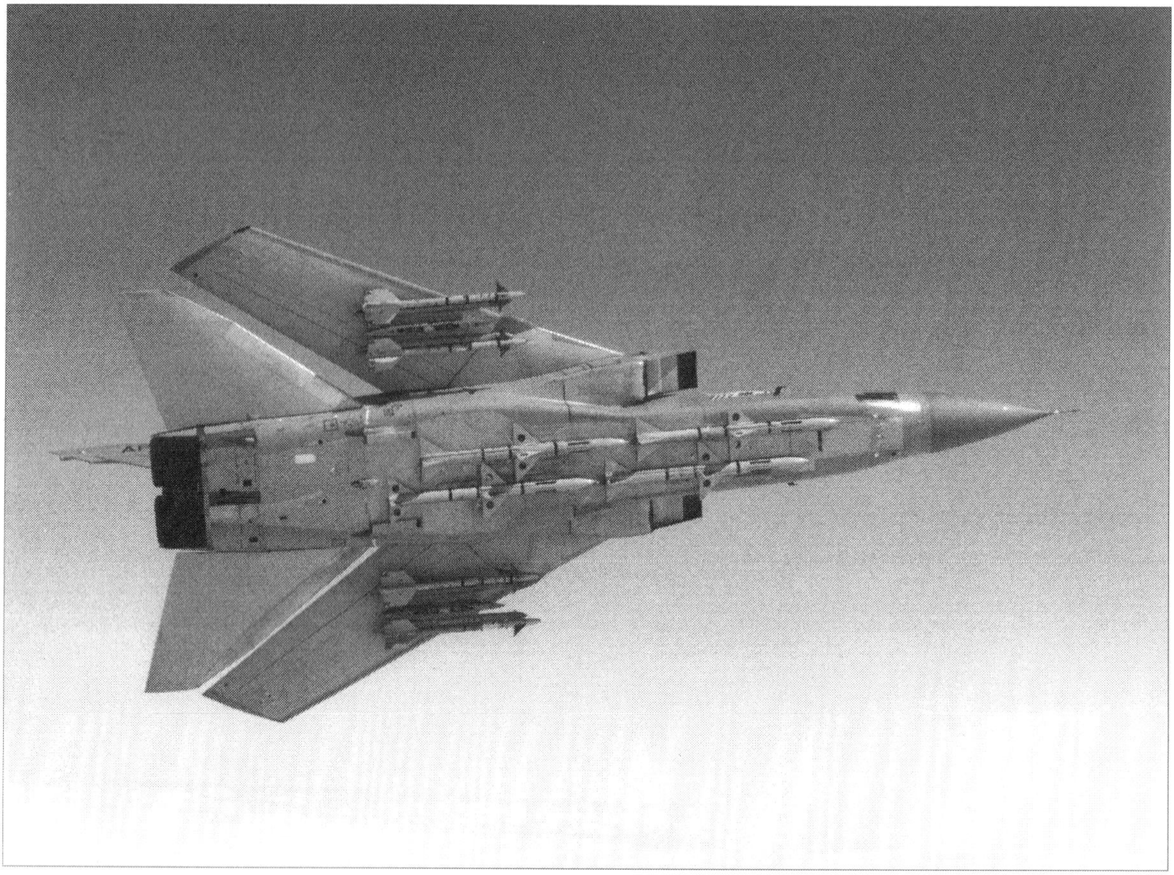

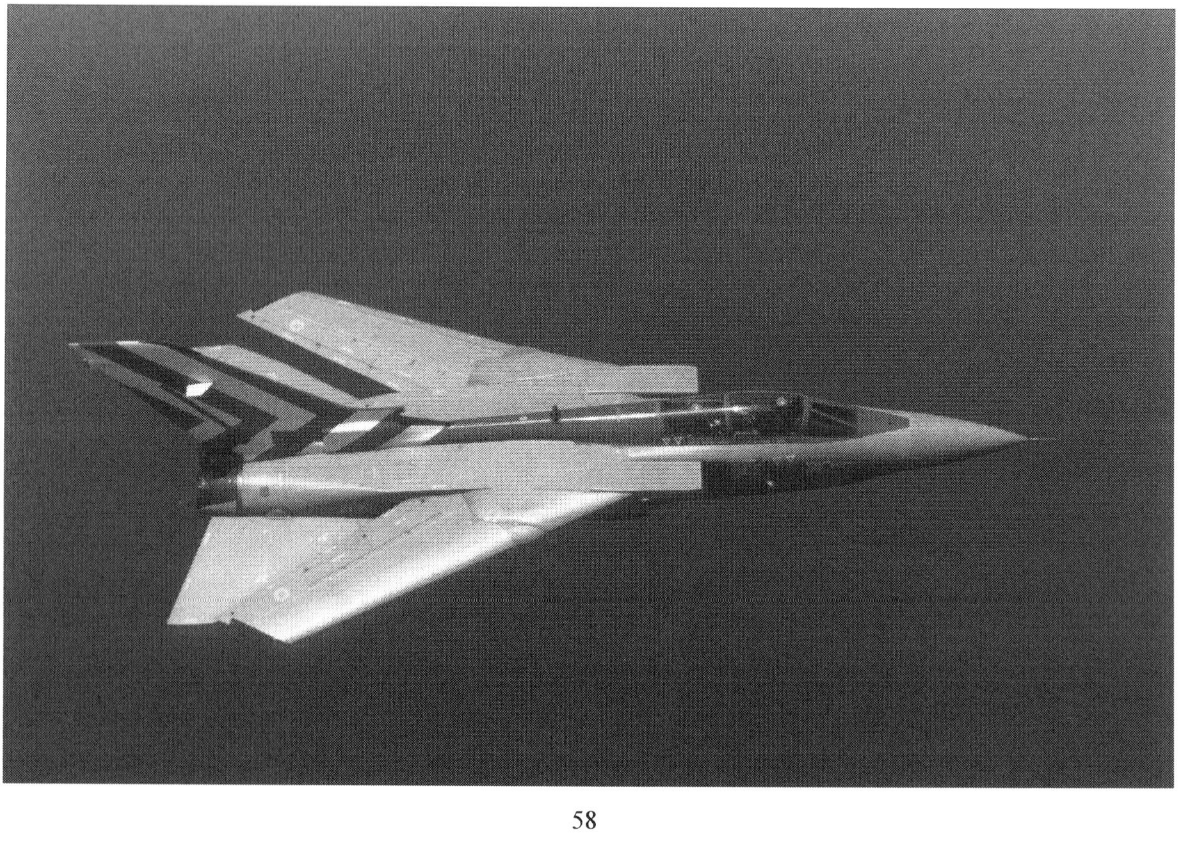

Previous page top: Wings in the maximum sweep back 67 degrees position, a Tornado F.3 from No.229 OCU reveals its underside showing the main armament of four under fuselage Skyflash BVR missiles and four AIM-9L Sidewinder SRAAM's. Previous page bottom: In 1990, No.229 OCU (No.65R squadron) displayed Tornado F.3 ZE907 in a special paint scheme of red spine and red and white vertical tail to commemorate the 50th anniversary of the Battle of Britain. Both BAE Systems

This page: Tornado F.3 ZE760 from No.229 OCU trails vapour from its wing tips as it pulls up during a training sortie. BAE Systems

No.23 squadron Tornado F.3s, made the first live interception of a Soviet aircraft - actually two Soviet Tu-95 'Bear' Ds - of a Leeming based Tornado F.3 over the Norwegian Sea on 10 November 1989. No.23 squadron commenced its work up on the F.3 on 4 January 1989 and was declared operational and assigned to SACUER on 1 August 1989. In January 1990, the Leeming wing Tornado F.3's took over the Northern QRA from the Leuchars based McDonnell Douglas Phantom FG.1's of No.43 and 111 squadrons allowing the Leuchars wing to prepare for re-equipment with the Tornado F.3. The Northern QRA commitment was handed back to Leuchars in January 1992.

The third and final Tornado F.3 squadron for the Leeming Wing was No.25 squadron, which disbanded as a Bloodhound surface to air missile unit on 1 August 1989, reforming the same day with Tornado F.3's at Leeming. The first Tornado for the squadron had arrived at Leeming almost a year previously, when ZE858 was delivered on 15 December 1988.

Tornado F.3 ZE202 from No.229 OCU (No.65 (R) squadron) lands at RAF Akrotiri, Cyprus in 1990. Tornado F.3 squadrons regularly deploy to Akrotiri for Armament Practice Camp. BAE Systems

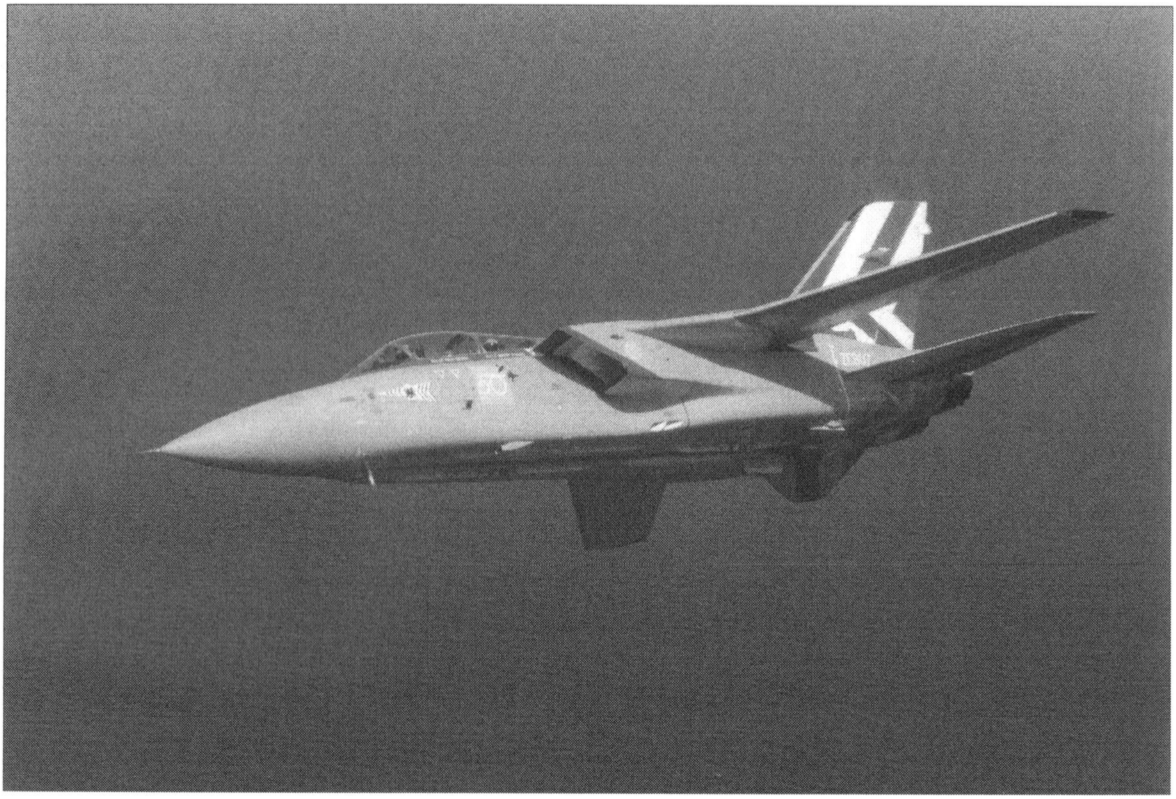
50

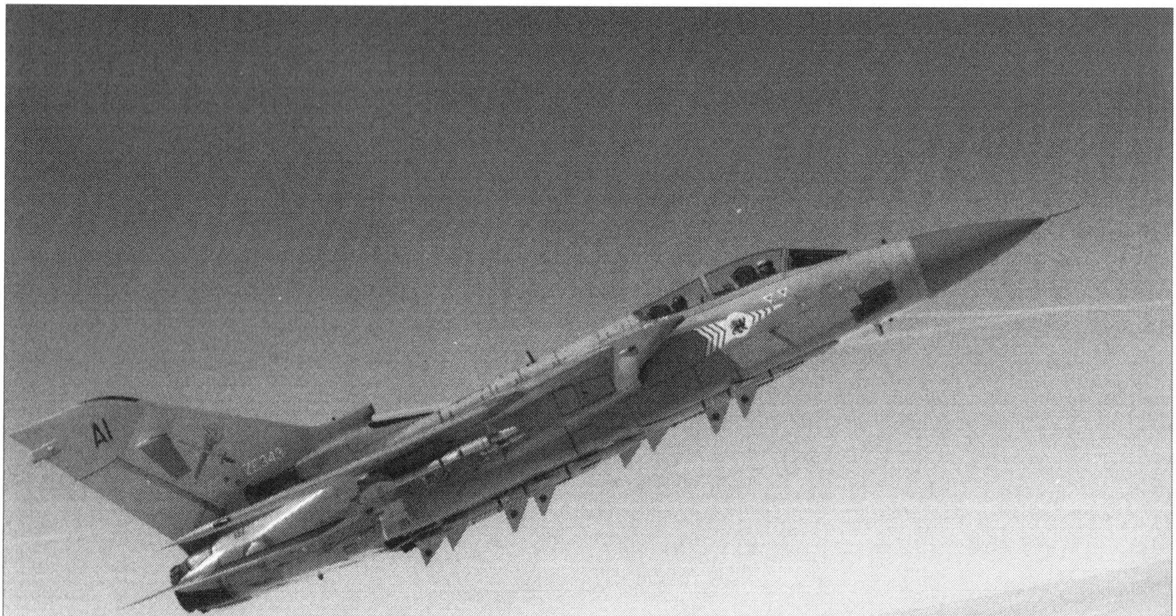
AI

Previous page top: No.229 OCU Tornado F.3 ZE907 resplendent in its 50th anniversary of the Battle of Britain colour scheme in 1990. Previous page bottom: No.220 OCU Tornado F.3 ZE343 during a training flight in the early 1990's. This page above: A No.229 OCU Tornado F.3 banks right for the camera showing its full complement of Skyflash and Sidewinder missiles. BAE Systems

In 1988, a quartet of No.29 squadron Tornado F.3's conducted a goodwill tour of the Far East. Tornado F.3 ZE759 is seen flying over the modern Bridge on the River Kwia top and the same aircraft above flies over Sydney Harbour with the world famous Sydney Opera House in the background. Both BAE Systems

A pair of No.5 squadron Tornado F.3's is seen flying in formation in 1991. The aircraft are in clean configuration with no external stores. BAE Systems

The first of the two Tornado F.3 squadrons which made up the Leuchars fighter wing was No.43 squadron 'Fighting Cocks'. The first F.3 for the Leuchars Wing was ZE963, which arrived on 23 August 1989 and was, used for ground familiarisation. The first two F.3's for No.43 squadron arrived at Leuchars very publicly during the annual Battle of Britain at Home day air show on 23 September 1989.

The last of the UK's planned F.3 squadrons; No.111 squadron completed the Leuchars Wing. No.111 was the last squadron to fly the Phantom FG.1 before the type was retired. The squadron re-formed as a Tornado F.3 unit on 1 May 1990 and officially declared to NATO SACUER, on 1 January 1991, although with No.43 Squadron deployed to the Gulf, No.111 squadron held the stations QRA commitment during December 1990.

On 6 June 1992, 3 RAF Tornado F.3's, ZE209, ZE790 and ZE792 departed RAF Coningsby for RAF Mount Pleasant in the Falkland Islands. The aircraft replaced the Phantom FGR.2's of 1435 Flight. The RAF has kept a handful of F.3's based

on the Falklands as a deterrent to any invasion plans Argentina may have. Although Argentina is unlikely to invade the Islands again, at least in the foreseeable future, the South American State has not relinquished its claim to the Islands.

Tornado F.3 ZE207 from No.29 squadron flies alongside a Jindivik target drone during a missile practice camp. The Jindivik, itself is not the target, but tows a small target behind, which is the target for the Tornado's missiles. BAE Systems

A pair of No.5 squadron Tornado F.3's, ZE254 (background) and ZG730 (foreground) fly in formation with a newly delivered RAF Boeing E-3D AEW (Airborne Early Warning) MK1 aircraft of No.8 squadron based at RAF Wadington in 1991. The introduction of the E-3D, which replaced the obsolete Shackelton AEW.MK3, which had previously equipped the RAF, revolutionised the capability of the UK's air defence forces. BAE Systems

Top: Leeming based No.11 squadron Tornado F.3 ZE200 launches an AIM-9L Sidewinder infrared guided short range air to air missile from the port outer stub station on the port inboard wing station in 1993. BAE Systems **Above: A No.11 squadron Tornado F.3 ZE968 is seen at RAF Leuchars in 1996.** Author

Above: From 2003, No.11 squadron took on the suppression of enemy air defence role armed with the MBDA ALARM 2 (Air Launched Anti-Radiation Missile). G H Lee
Left: The Leeming Tornado Wing in late 2004 consisted of No.11 squadron (foreground) and No.25 squadron (background). Both Tornados' are armed with a single Skyflash and a pair of Sidewinders. Crown Copyright

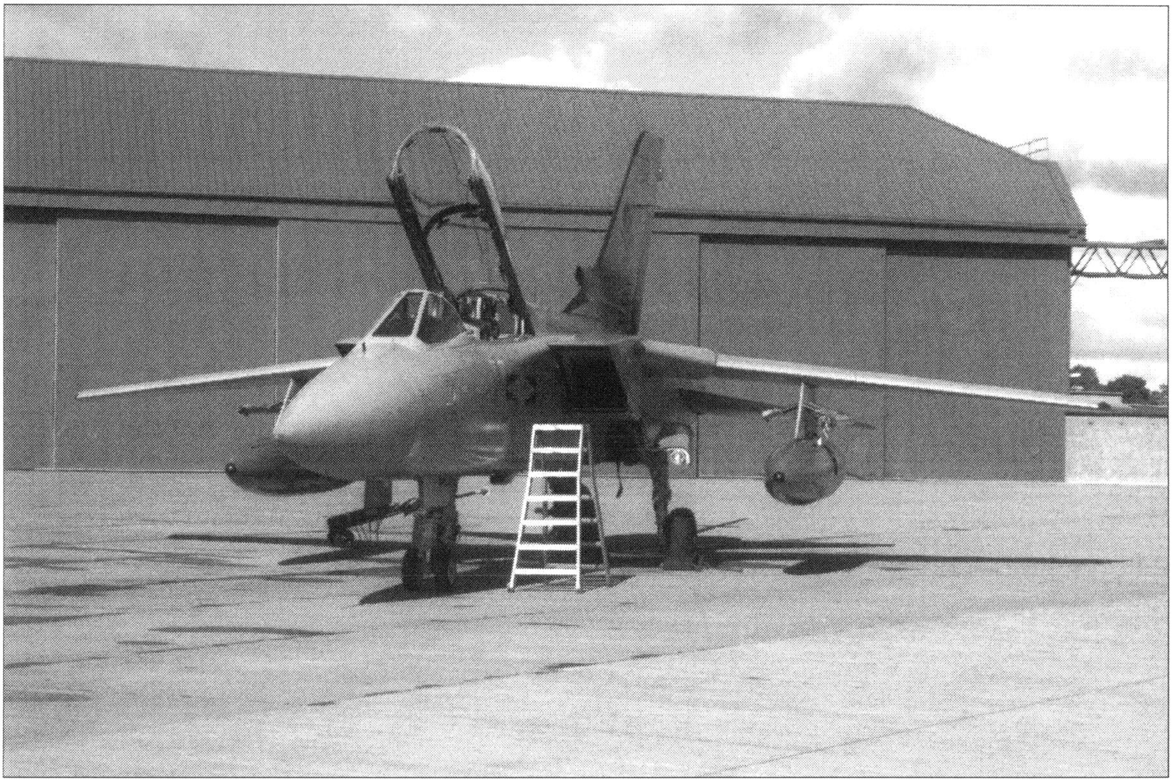

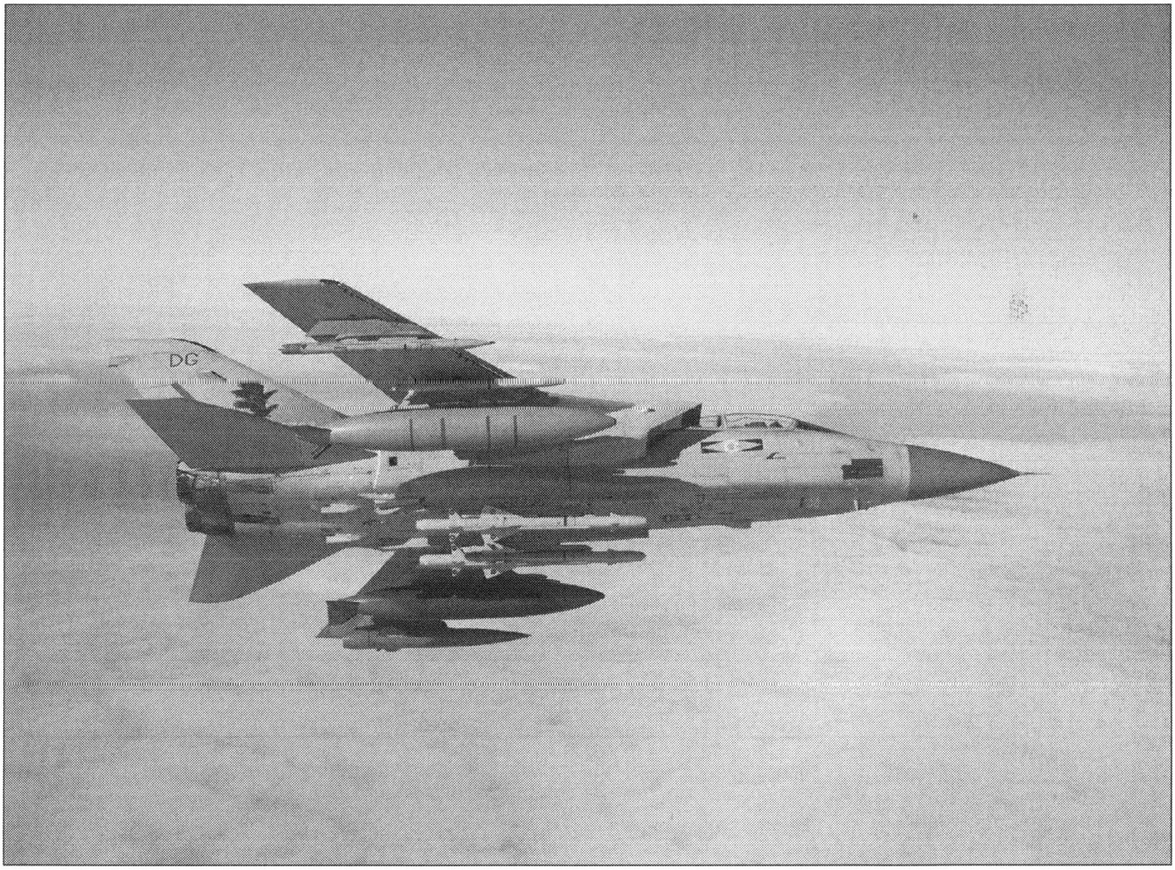
DG

Previous page top: A No.11 Squadron Tornado F.3 sits at the ramp at RAF Leuchars in March 1999. The aircraft is configured with the smaller 1500-litre external fuel tanks normally seen on the Tornado IDS variant. Author Previous page bottom and this page: No.11 squadron Tornado F.3 ZE763 was one of a batch converted to conduct the suppression of enemy air defence role armed with the MBDA ALARM 2. The introduction of an air to surface role sees the Tornado F.3 at last to begin to show some of the true MRCA potential always present in the design. Both G H Lee

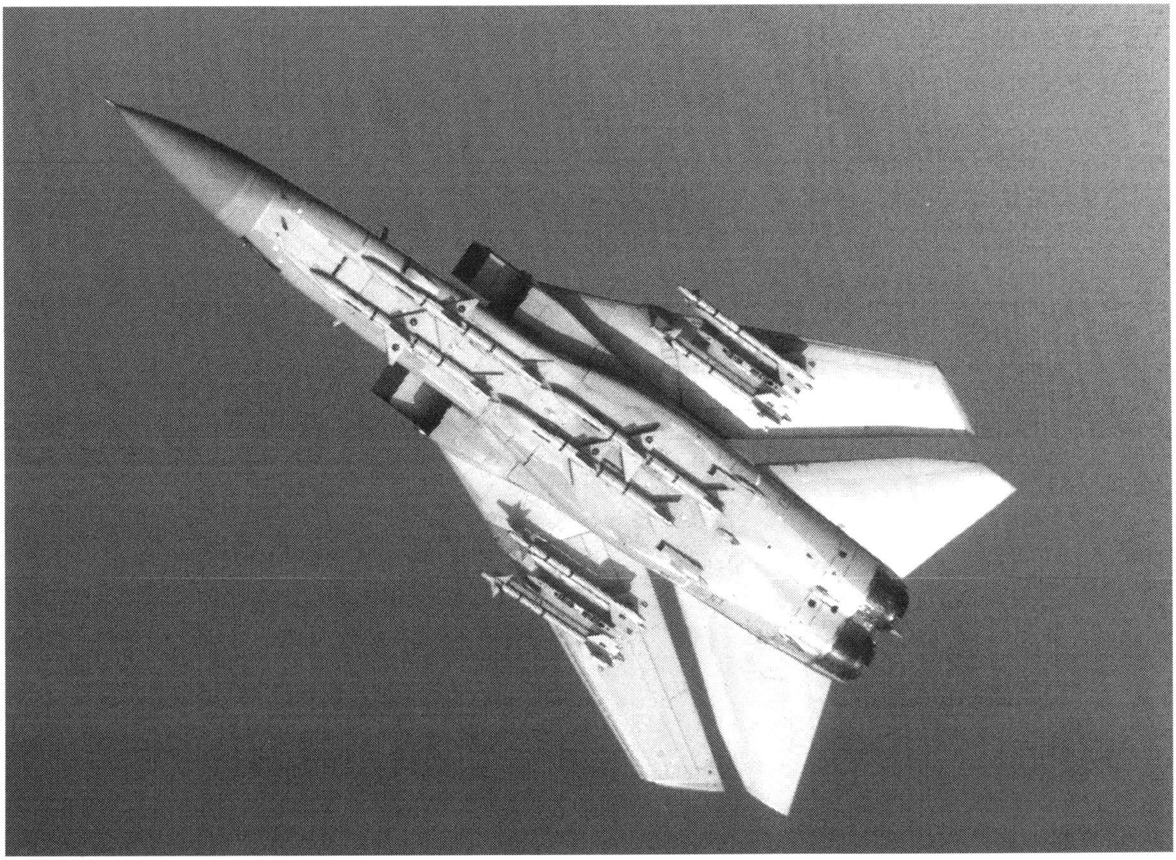

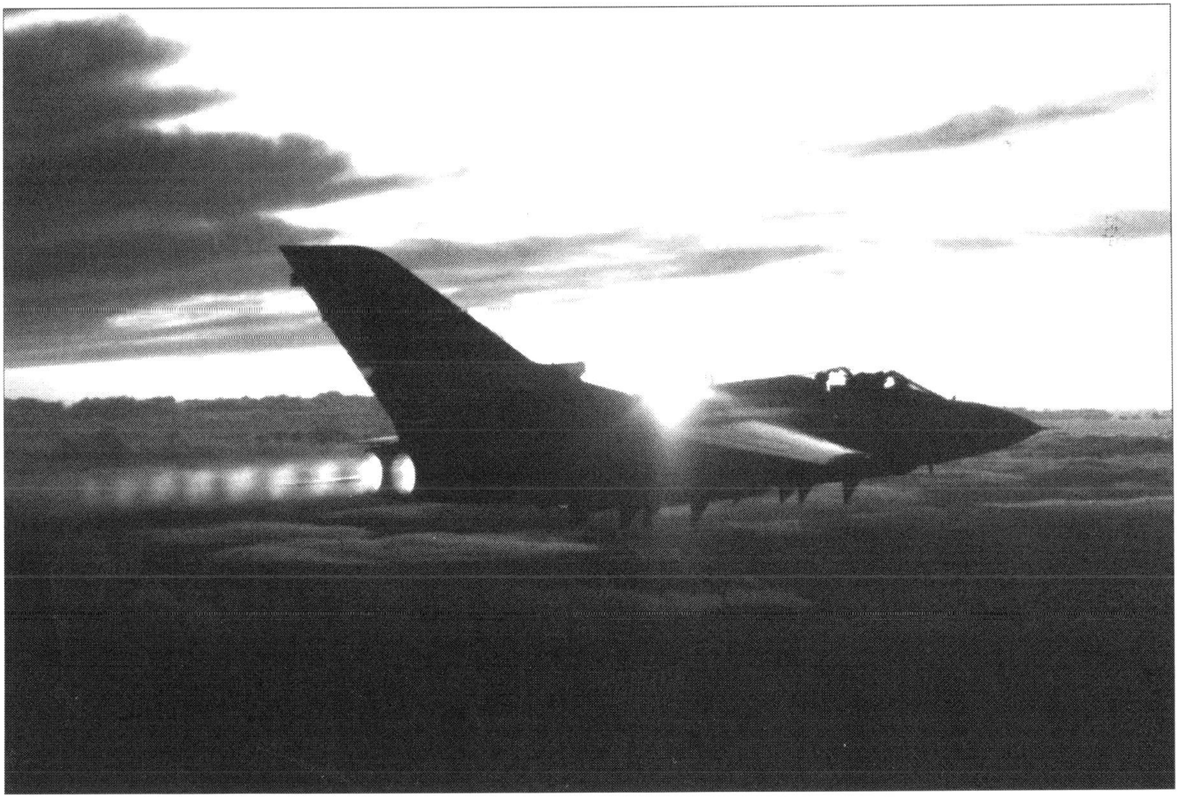

Previous page top: An early production Tornado F.3 shows its underside. Previous page bottom: A Tornado F.3 in afterburner is silhouetted against the evening Sun. This page right: A Tornado F.3, ZE160 (EX) from No.23 Squadron is seen at RAF Leeming in 1992. Author **Above**: No.25 Squadron was the third of Leemings squadrons to form on the **Tornado F.3.** Crown Copyright

Top: A Tornado F.3 from No.25 Squadron is seen during a training flight in the early 1990's. BAE Systems **Left: This** portrait of a No.25 squadron underside shows to advantage the under fuselage carriage of Skyflash medium range air to air missiles. One of the forward missile recesses is empty showing the indent allowing the semi-recessed carriage of the forward pair of missiles. Crown Copyright

Above: Tornado F.3 ZE737 FF from Leeming based No.25 Squadron passes a No.111 squadron Hardened Aircraft Shelter as it begins its take-off run at RAF Leuchars in September 1996. Top: Tornado F.3 ZE936 DL from No.11 squadron and ZE737 FF from No.25 Squadron sit on the ramp at RAF Leuchars in September 1996. Both Author

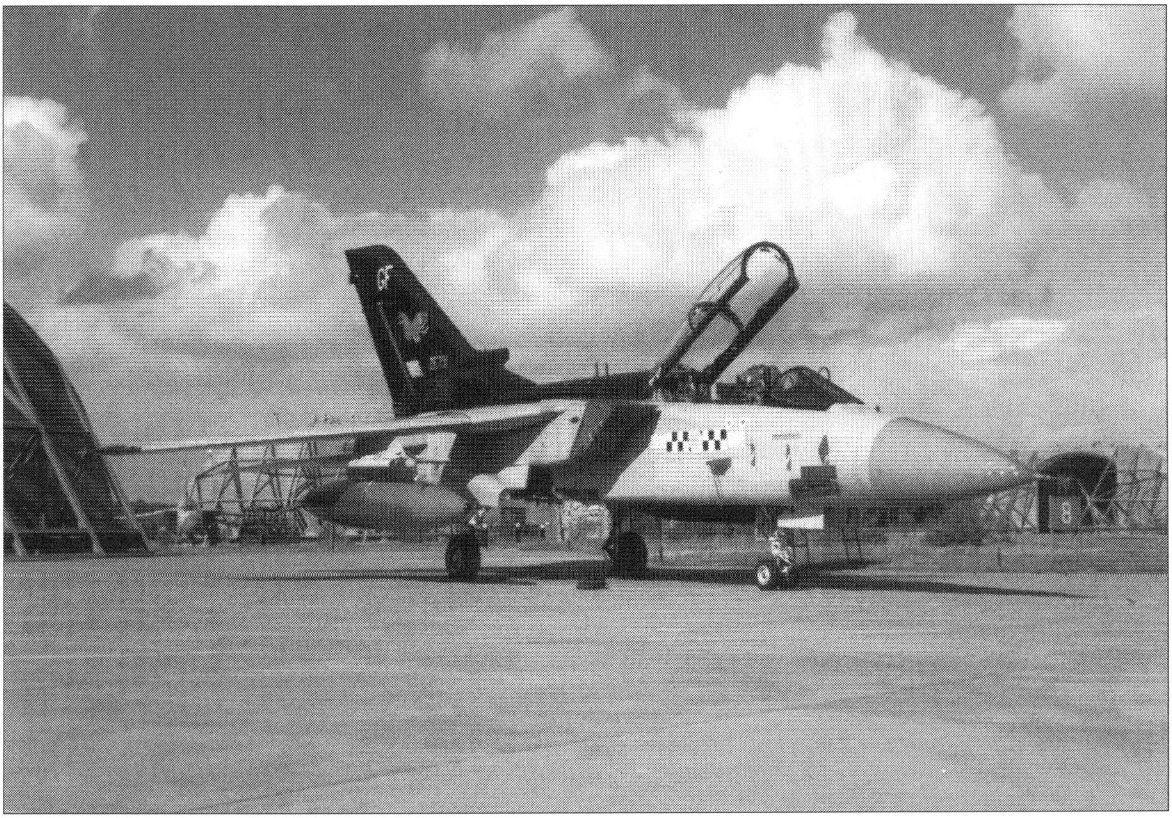
GF

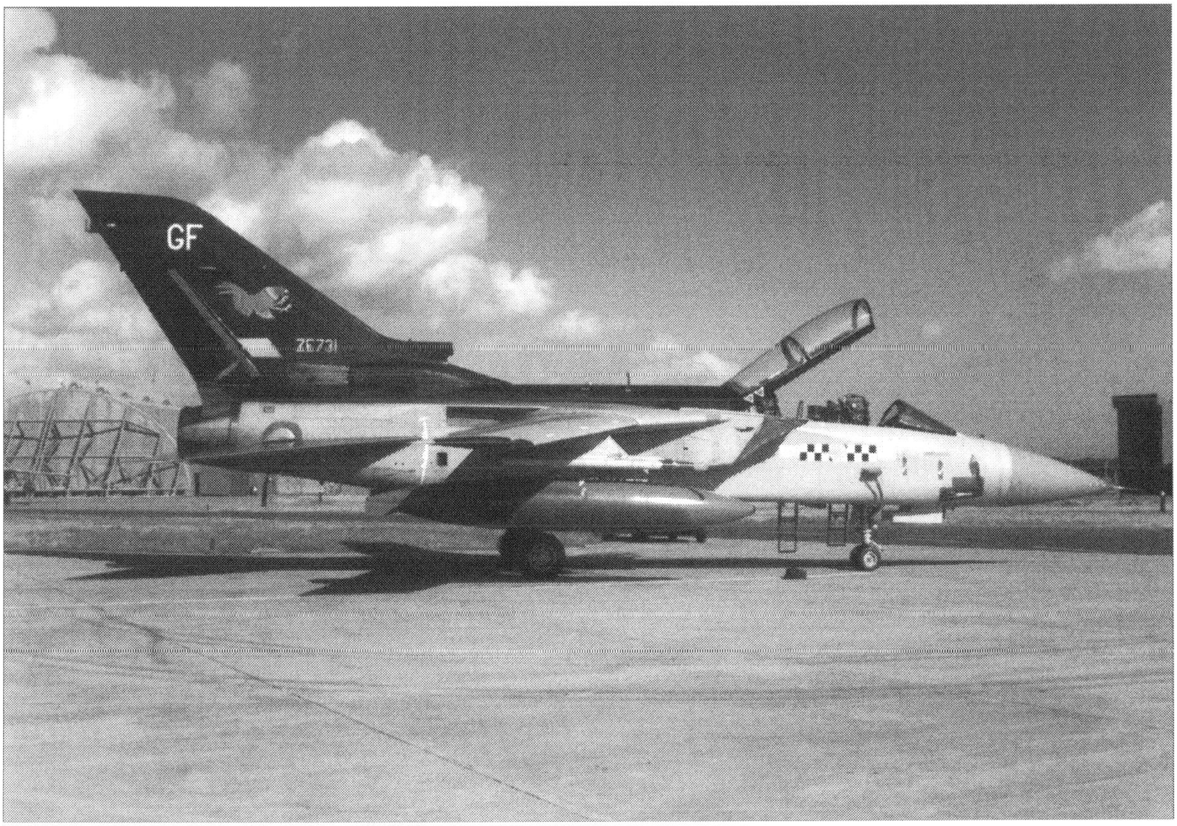
GF
ZE731

Previous page top and bottom: In 1996, No.43(F) Squadron painted Tornado F.3 ZE731 GF in a special scheme to commemorate the 80th anniversary of the formation of the squadron. Both Author This page top: A No.43(F) Squadron Tornado F.3 sits in the 'Q' Hardened Aircraft Shelter at RAF Leuchars in September 1993. Above: On 17 September 1993, ZE730 was scrambled from RAF Leuchars on only the second QRA launch in two-years. The aircraft is seen returning to Leuchars. The Russian aircraft were turned back by Norwegian F-16's before the F.3's reached them. Both Author

Top: Seen in No.43(F) Squadron markings at RAF Leuchars in 1993, ZE733 has its airbrakes - one each side of the vertical tail – deployed. Above: No.43(F) squadron Tornado F.3 ZE296 GR sits on the ramp at RAF Leuchars in the late 1990's. Both Author

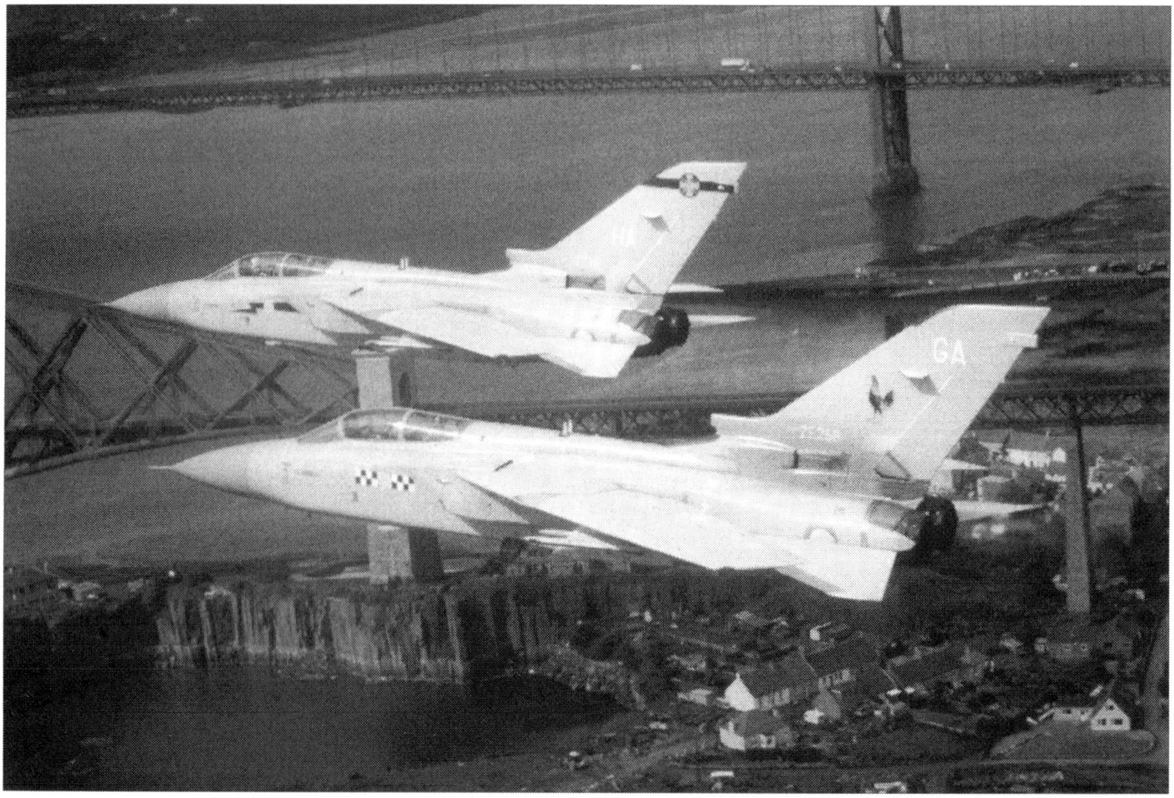
HA
GA

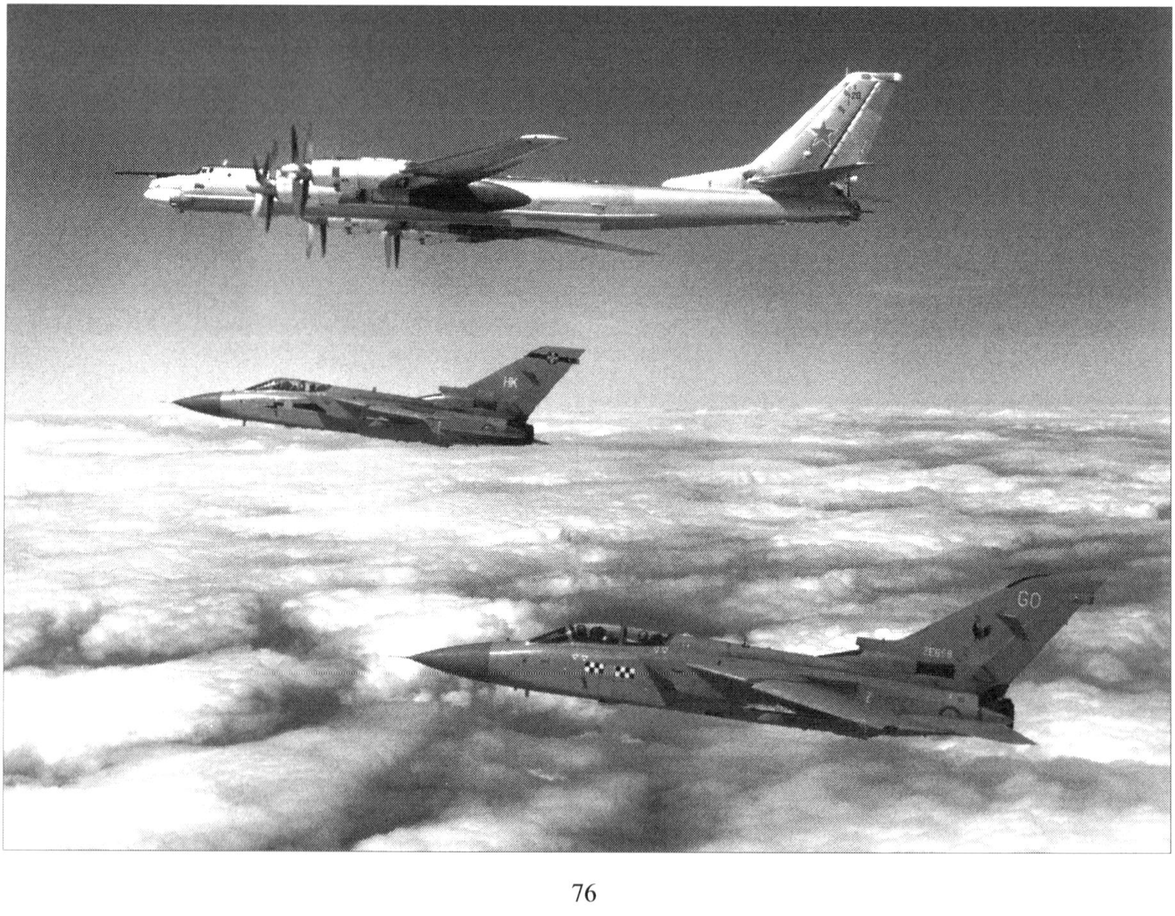
HK
GO

Previous page top: A No.43(F) Squadron Tornado F.3 ZE258 flies in formation with a No.111(F) squadron aircraft ZE736 over the Forth road and Rail bridges in the early 1990's. Previous page bottom: The Tornado F.3 in its element! A pair of Leuchars wing Tornado F.3's escort a Russian Tu-142 'Bear' into UK airspace in July 1993, for that years Royal International Air Tattoo. ZE835 HK (background) is from No.111(F) Squadron and ZE858 GO (foreground) is from No.43(F) Squadron. Both Crown Copyright This page top: ZE292 from No.11(F) Squadron taxis out from a HAS at Leuchars in 1993. This page above: No.111(F) Squadron decorated ZG776 in an attractive paint scheme to celebrate the unit's 75th anniversary. Both Author

Top: A pair of No.111(F) squadron aircraft flanks A No.25(F) Squadron Tornado F.3. The aircraft nearest camera is in clean configuration. The centre aircraft has a single MBDA Skyflash on one of the rear fuselage stations, while the farthest aircraft has a single Skyflash on one of the forward fuselage missile recesses. Above: ZE837 from No.111(F) Squadron taxis to the units HAS complex following a sortie in September 1993. Crown Copyright and Author

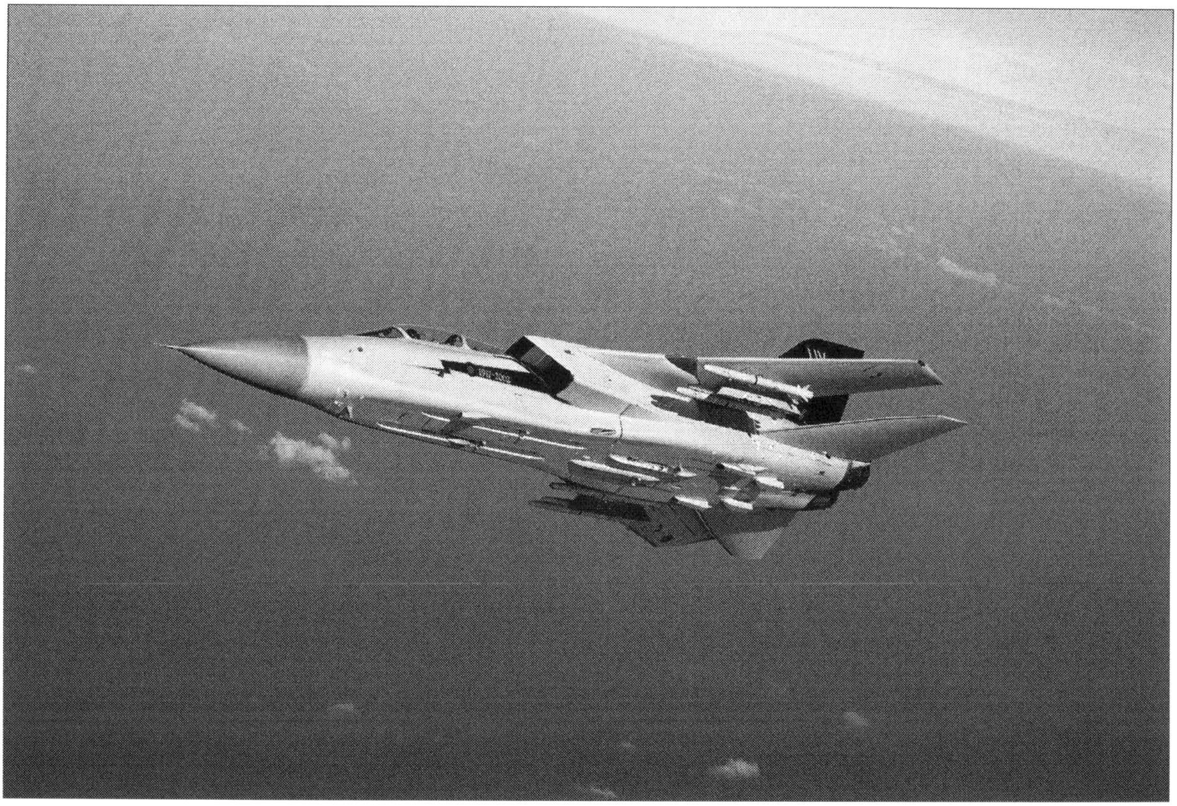

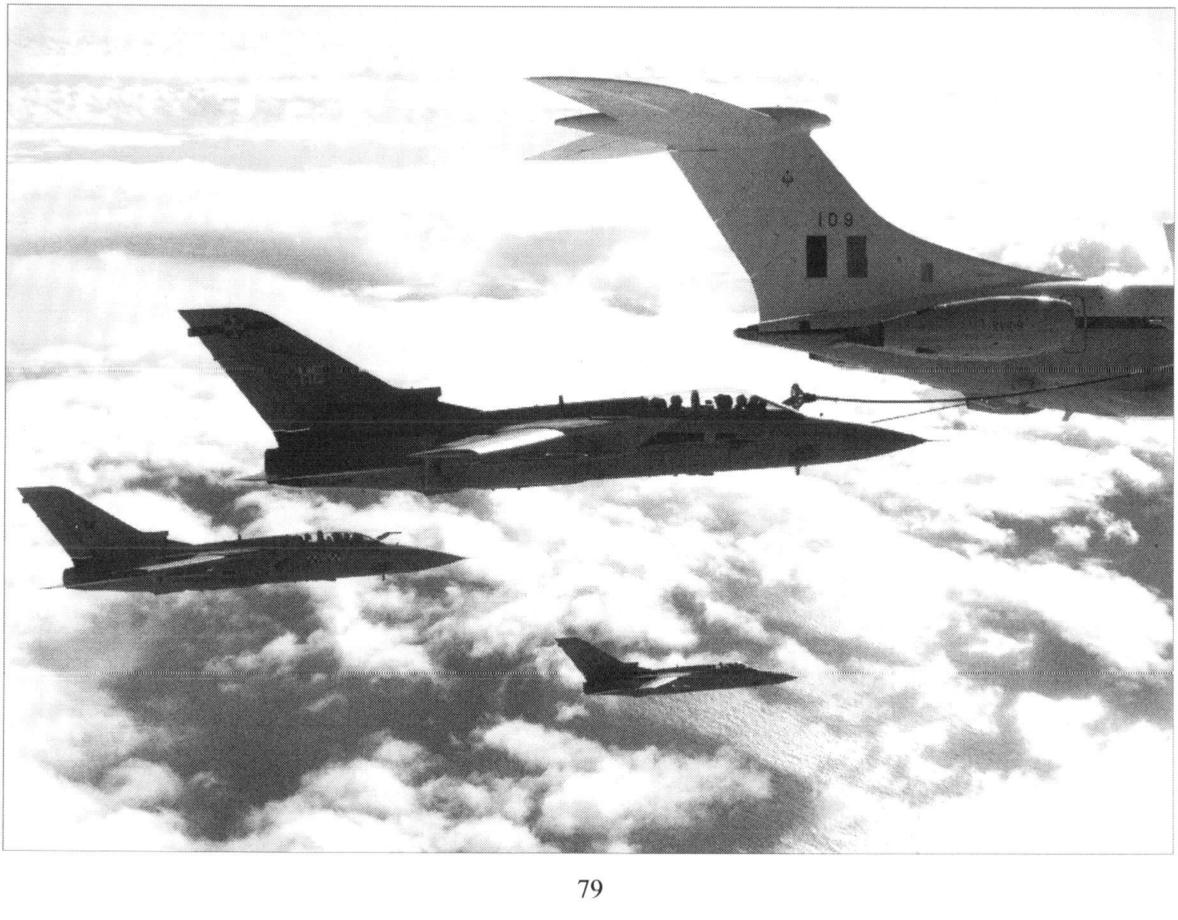
109

Previous page top: A Treble One squadron Tornado F.3 banks for the camera revealing its MBDA ASRAAM missiles. ASRAAM entered RAF service in 2002. Previous page bottom: ZE156 from Treble One squadron and a pair of No.43(F) squadron Tornado F.3's refuel from a RAF VC 10 Tanker. G H Lee and Crown Copyright This page: A No.65 (Reserve) squadron Tornado F.3 carrying a full load of Skyflash and Sidewinder missiles goes into an almost vertical climb. BAE Systems

Top and above: Various formations of Tornado F.3, McDonnell Douglas Phantom FGR.2 and Boeing E-3D Sentry AEW MK1's were flown to mark the disbanding of No.56 squadron with Phantom FGR.2's and re-formation as No.56(Reserve) squadron equipped with Tornado F MK3's. Both BAE Systems

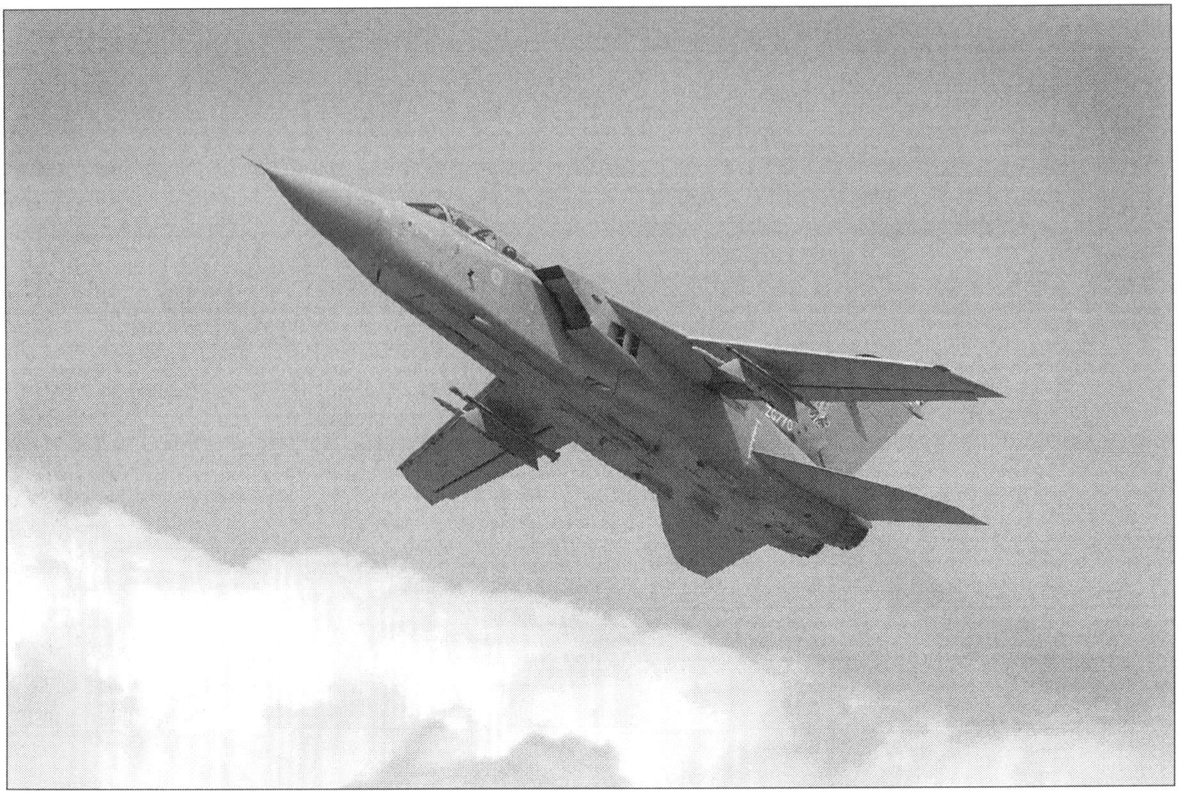

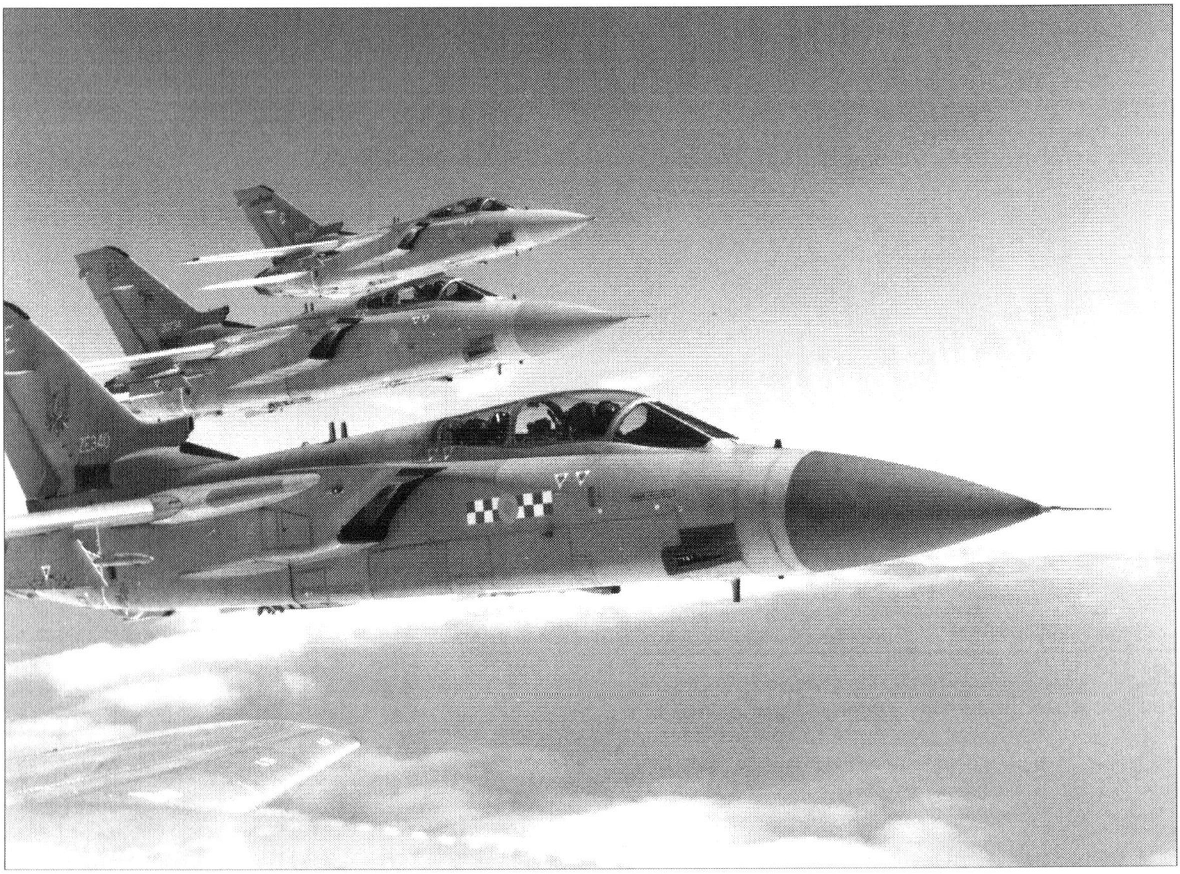
ZE340

ZE340 (foreground) flies in close formation with ZG734 from No.29 squadron (centre) and ZH555 from No.5 squadron (background). Crown Copyright This page top: ZE882 from the Tornado F.3 OEU flies in formation with a RAF Boeing E-3D Sentry AEW MK1 from No.8 squadron. This page left: A No.56(Reserve) Squadron Tornado F.3 banks left with the wings in the intermediate sweep position. BAE Systems and Crown Copyright

Previous page top: Tornado F.3 ZG770 in flight in clean configuration. Rolls Royce Previous page bottom: No.56(Reserve) squadron Tornado F.3

Left: A Tornado F.3 from the Falklands based 1435 Flight flies close to the camera aircraft. Tornado F.3's serving in the Falkland Islands normally fly armed with only a pair of Skyflash beyond visual range air to air missiles in the forward fuselage recesses and the usual complement of AIM-9L Sidewinder short range air to air missiles on the inboard wing stub

Top: In 1992, three Tornado F. MK3's were delivered to 1435 Flight at RAF Mount Pleasant in the Falkland Islands in the South Atlantic as replacements for the Flights McDonnell Douglas Phantom FGR.2's.

pylons. Both Crown Copyright

From top left to bottom left: Units operating the Tornado F.3 in early 2004 included No.11(F) and 25(F) Squadrons at Leeming and No.43(F) and 111(F) Squadrons at Leuchars. Author Crown Copyright and both Author Top right: No.56(Reserve) Squadron was also operating from Leuchars; ZE163 among its aircraft on strength. Author Centre right: Tornado OEU operated from Waddington in early 2004 before a scheduled return to Coningsby. Crown Copyright Above: The only permanent non-UK-deployed Tornado F.3 unit was No.1435 Flight based in the Falkland Islands. Crown Copyright

Top left and centre top left: In 1996, No.43(F) Squadron celebrated the 80th anniversary of the formation of the unit as part of the Royal Flying Corp in 1916. The squadron painted ZE731 in a special eye catching scheme of black tail and spine and what can only be described as a mythical predatory Cock (male chicken) complete with claws on the tail. Both Author

Bottom left and centre bottom left: In 2002, No.111(F) Squadron celebrated the unit's 85th anniversary of its formation as part of the Royal Flying Corp in 1917. The unit painted ZE159 in a special scheme to commemorate the occasion. The vertical tail was painted black adorned with a large yellow cross with crossed swords in red. Both Author

From the early 2000's, RAF Tornado F.3 squadrons began to adopt toned down unit insignia on the tails of their aircraft as shown on ZE158 from No.111(F) Squadron top right and ZE838 on No.43(F) Squadron above. Both Author

The RAF received the last of 152 production Tornado F MK3's in 1993 when production of the ADV (Air Defence Variant) ceased. Production for the UK also included 3 prototypes and 18 Tornado F MK2's. The Royal Saudi Air Force received 24 Tornado ADV's bringing total ADV production to 197 aircraft. Here the last production ADV, Tornado F MK3 ZH559 tucks-up its undercarriage during take-off from BAe Warton for delivery to the RAF in March 1993. BAE Systems

In 1992, No.56 (Reserve) squadron took over from No.65 (Reserve) squadron as the Tornado F.3 OCU at RAF Coningsby after retiring its Phantoms. On 27 March 2003, the 11 Tornado F.3's from No56(R) squadron moved from Coningsby to RAF Leuchars in Fife, Scotland to allow Coningsby to be renovated ahead of formation of the first RAF Eurofighter Typhoon squadrons in 2003/4. Leuchars has only 22 HAS's, used to house the 13 Tornado F.3's of No.43(F) squadron and 12 F.3's of No.111(F) squadron, therefore, No.65(R) squadron operates from hanger complexes used in Leuchars pre-HAS days. The Tornado F.3 OEU (Operational Evaluation Unit), which was also based at Coningsby was moved to

RAF Waddington and was scheduled to return to Coningsby in April 2004 where it was to be joined by the Strike OEU, combining to form the Fast Jet OEU.

A pair of No.111(F) squadron Tornado F.3's sit on the wing of a RAF VC10 in-flight refuelling tanker aircraft as a pair of French Air Force Mirage 2000D strike aircraft take on-fuel from the tankers two wing hose and drogue units. Crown Copyright

The penultimate Tornado F.3 for the RAF was ZH558 seen operating with No.43(F) squadron at RAF Leuchars during exercise Elder Joust in September 1993. Author

As well as the Tornado F.3 OEU (Operational Evaluation Unit, a number of test agencies have operated the Tornado ADV prototypes as well as Tornado F.2 and Tornado F.3 production aircraft. The DRA Farnborough operated Tornado F.2(T) ZD902 as the TIARA (Tornado Integrated Avionics Research Aircraft), which was equipped with a GEC Marconi (now BAE Systems) Blue Vixen multi-mode pulse-Doppler radar installed in place of the same companies Foxhunter. The Blue Vixen was developed for the BAE Systems Sea Harrier FA.2 STOVL (Short Take-Off and Vertical Landing) carrier strike fighter in service with the UK Royal Navy Fleet Air Arm. In the TIARA, the Blue Vixen was integrated with a GEC Sensors IRST (Infra-Red Search and Track) system. The aircraft's front cockpit was modified to simulate an advanced single-seat fighter cockpit with a heliographic HUD (Heads-Up Display) and three full colour multi-function display screens. HOTAS controls and a JTIDS terminal were installed, these also being adopted by the operational Tornado F.3 fleet. The TIARA Tornado also had the ability to use a HMSS (Helmet Mounted Sighting System). Externally, the aircraft carried modified under fuselage fuel tanks, which housed an array of test equipment and sensors. The modified fuel tanks could be fitted with either high or low speed nose cones. When displayed at the 1994 Farnborough show, the TIARA Tornado was fitted with modified fuel tanks with an optical window apparently for a FLIR (Forward Looking InfraRed) system. The TIARA Tornado moved to Bascombe Down in the mid-1990's with the rest of the DRA aircraft fleet an is now operated by Quintiq, the private sector part of the former DERA

A Tornado F.3 sits in a Hardened Aircraft Shelter at Leuchars in 1993. Author

Tornado F3 ZE160 DV from No.11(F) squadron in flight with wings in the maximum sweep position.
Crown Copyright

The first and third Tornado ADV prototypes, A01 (ZA254) and A03 (ZA283) and a pair of production Tornado F.2's, ZD899 and ZD939 were involved in AI.24 Foxhunter radar development flying test flights from BAe Warton in Lancashire. The main target aircraft for Foxhunter development flights were initially a handful of BAe Lightning F.3 and F.6 fighters, which had been retired from operational service in the 1980's, although other aircraft were used including other Tornado's. One objective of the radar development flights was to demonstrate the Foxhunter radar ability to detect and track one or more target aircraft in a multitude of engagement scenarios at high, medium or low altitudes, at various speeds and in a high clutter electronic counter countermeasures environment.

The second Tornado ADV prototype, A02 (ZA267) and the second production Tornado F.2 (ZD900) were for a while operated by the A&AEE (Aircraft & Armament Experimental Establishment) at Bascombe Down as trials aircraft. ZA267 was retired and delivered to RAF Marham as a ground instruction airframe on 11 December 1998. The aircraft was being disposed of by Bascombe Down, but was delivered to the Tornado Maintenance School as replacement for Tornado GR.1 ZA564, which was returned to service. At the time of its retirement ZA267 was the last of the three prototypes still flying.

The defence of the UKADR (UK Air Defence Region), in peacetime is carried out in an operation known as the Quick Reaction Alert (QRA) or 'Q' for short. RAF Leuchars is the only fighter station, which has a permanent 24 hours a day, 365 days a year 'Q' commitment. The 'Q' is organised in such a way that the duty squadron at the time is responsible for providing both aircraft and aircrew for Q1 and Q3, while the non-duty squadron is responsible for aircraft and aircrew for Q2. Q1 and Q2 are held at 15-minutes alert whereas Q3 is on 60-minutes alert. For QRA duty each Tornado F.3 was armed with four Skyflash medium range air-to-air missiles (AAM) and two AIM-9L/M Sidewinder air-to-air missiles, along with the built-in 27-rnm Mauser cannon. In addition the aircraft carry two 2250-litre external fuel tanks. This configuration is known as 'Lima fit'. With AMRAAM and ASRAAM introduced to the inventory, these weapons can also be used to arm 'Q' Tornado's, with the ASRAAM in particular being used in preference to the much less capable AIM-9L/M. From 2004, AMRAAM equipped Tornado F.3's from No.43(F) and 111(F) squadrons were conducting QRA operations armed with AMRAAM and ASRAAM.

Tornado F.2, ZD902 was operated by the DRA, DERA and later Quintiq. The aircraft was equipped with a number of advanced sensors and equipment including the Blue Vixen multi-mode pulse-Doppler radar developed for the Sea Harrier. Quintiq

Above: A No.111(F) squadron F.3 sits in a HAS at RAF Leuchars in the mid-1990's. Author

With the introduction of the Hardened Aircraft Shelter (HAS) the preferred method is to operate directly from the shelters, as opposed to the purpose built 'Q' sheds of days gone by. No.111 squadron's shelters are used regardless of which squadron holds 'Q' duty. The reason for this is that No.111's shelter complex is closer to the main runway 09/27. Normally Q1 and Q3 were housed in the same shelter, with Q3 sitting behind Q1. Aircrew for both are available on site and -serve a 24-hour -period of duty.

The 'Q' Tornado's are required to investigate and identify any aircraft entering the UKADR, without filing a flight plan. Main trade has typically been long range Soviet and now Russian Tu-95/142 'Bear' and Tu-16 'Badger' reconnaissance aircraft, although a few unidentified civil aircraft have also been intercepted. Prior to the early 1990's, live QRA launches were an almost daily practice. Today they are much less frequent due to the break up of the former Soviet Union and the more friendly relations enjoyed between NATO and Russia. In the early 1990's, Leuchars Tornado F.3's became more used to escorting former

Commonwealth of Independent States (CIS) and Russian aircraft visiting the station. These have included the Russian Knights aerobatics display teams with their 6 Sukhoi Su-27 'Flanker' fighters and Ilyushin IL-76 'Candid'. The team performed their first ever full public display at Leuchars on 21 September 1991. The Russians returned in 1992 with a pair of Su-27's and a Su-24MR 'Fencer' E, the first time a 'Fencer' had visited a front line RAF station.

ZE338 HG from No.111(F) squadron is towed from the squadrons HAS complex to the maintenance hanger at Leuchars. Author

Top: ZG730 GL from No.43(F) squadron is scrambled on a QRA on 17 September 1993. The two Russian Aircraft were turned back before the F.3 arrived on station. Above: Tornado F.3 ZE762 GM is towed from No.43 squadrons HAS complex to No.111 squadrons HAS complex for QRA duty. Both Author

On 6 September 1991, Tornado F3s from Leuchars were scrambled on the last QRA alert for a period of two years. The sleepy two-year period ended on 7 September 1993, closely followed by another scramble on 17 September 1993. On both occasions the aircraft, a 'Cub' and a 'Coot', were turned back by Royal Norwegian Air Force Lockheed Martin F-16A Fighting Falcons before the Tornado F.3s arrived on station.

The QRA alert system would be replaced in times of crisis or war by aircraft on Combat Air Patrol (CAP). It is in this type of operation that the Tornado F.3 comes into its own with few equals. A typical CAP would involve two aircraft, fully armed with four Skyflash or AMRAAM and 4 AIM-9L/M or ASRAAM's and a pair of 2250-litre external fuel tanks. The aircraft would fly a race track pattern about 300-miles from base for a

Top: ZE289 HF from No.111(F) squadron takes-off in full afterburner during a Tactical Leadership training exercise in 1995. Above: A No.111(F) squadron lands back at Leuchars following a sortie during a Tactical Leadership Training exercise in 1995. Both Author

duration of around 3-hours, after which they return to base or remain on station with the assistance of in-flight refuelling from RAF VC10 or Tristar in-flight refuelling tankers. With the combination of its powerful AI.24 Foxhunter radar and Skyflash or AMRAAM with their 30+ mile beyond visual range capability, the Tornado F.3 is one of, if not the most capable interceptor in any air force.

Most Tornado F.3 maintenance and servicing were carried out at station level, with first line servicing of the aircraft carried out at squadron level. Second line servicing (Primary star and Minor Star) was conducted by the

Top: A No.11 squadron Tornado F.3 sits on the ramp at RAF Leuchars in March 1999. Author Above: Like the Phantoms FG.1/FGR.2's before, them, the RAF

Tornado F.3 sometimes operated with Hawk T.1A's in the mixed fighter force role. BAE Systems

A pair of No.56(Reserve) Squadron Tornado F.3's fly in formation with an E-3D Sentry AEW 1 and a pair of No.56(R) Squadron McDonnell Douglas Phantom FGR.2's. BAE Systems

Tornado Aircraft Servicing Flight (TASF), most of whose time is taken up with minor servicing which is due every 125-hours involving about 4/5 days in the TASF hangar. During Primary Star servicing the aircraft receives a full wash down, although the rear fuselage and fin/rudder require more frequent washing because of the Tornado F.3 RB199 MK104 thrust reverse buckets, which deflect the exhaust gas forwards. Major servicing, which is required every 2,000-hours, is the responsibility of the Maintenance Unit (MU) at RAF St Athan. The General Engineering Flight is responsible for paint scheme touch-ups and application of squadron markings. Aircraft requiring a full repaint are normally sent to RAF St Athan.

ZE964 from No.11 squadron with refuelling probe extended at Leuchars in the early 1990's. Author

The Tornado Propulsion Flight (TPF) provides RB199 engines and support and conducts all deep strip and second level work. The unit's duties include providing a replacement engine for any F.3 requiring such. When a replacement engine is required the aircraft will be taken from the squadrons HAS complex to the TPF where a new engine will be fitted. For this a number of spare engines are stored on base. TPF facilities include an engine bay and test house with computer controlled running procedures along with a de-tuner, allowing engine running even at night without annoying noise pollution. Normally an engine will be removed and a replacement installed in its place in about eight hours, although it can be, and often is, done in much less than this. In emergencies it is possible to remove an engine in under an hour and fit its replacement in a similar time scale. This capability is a far cry from the days of the Lightning and Phantom when a similar job could take days.

In 1979, a decision was taken to arm the RAF Hawk T.1 with a pair of AIM-9L Sidewinder air-to-air missiles in order to give it a secondary war role of point defence of the UK's airfields. It had originally been hoped to form a third Lightning squadron but this idea was dropped in favour of the Hawk. The conversions were done under the name of the Hawk war role programme with the work being carried out by BAe. Work was started in early 1980, with the first Sidewinder firing from a Hawk taking place the same year. Due to some minor problems the Sidewinder Hawk combination was not cleared for use until May 1983. An official contract for the conversion of eighty-nine Hawk T MK1's to T MK1A standard was awarded to BAe.

The No.11 squadron unit emblem is shown just below the twin cockpit. Author

Tornado F.3's from No.43(F) Squadron taxi at Leuchars. ZH558 (top) is the CO's aircraft adorned with the fin tip flash. ZE858 above has deployed its thrust reverse buckets to aid braking. Both Author

Included in the modifications was the fitting of a Ferranti F.195 weapon sight along with strobe lights and a twin gyro platform to provide a high altitude accuracy reference system. The aircraft selected for modification came from the two Tactical Weapons Units and the Central Flying School (CFS) at RAF Scampton. This latter unit included the RAF Red Arrow aerobatics display team, which would have formed as an operational squadron in the event of war. The other units were No.19, 74, 92 and 208 reserve squadrons. The last of the modified Hawks was delivered to the RAF in August 1986, by which time the operational role had changed from that of point defence of military installations to the more aggressive one of meeting the enemy as far from the United Kingdom's shores as possible.

Top: ZE339 from No.25(F) squadron at Leeming was the 1992 Tornado F.3 display aircraft. Above: Tornado F.3 ZE161 FG from No.25(F) squadron taxis after a sortie in 1992. Both Author

As the Hawk had no radar of its own it would fly with Tornado F.3s as part of a Mixed Fighter Force (MFF). Typically, a single Tornado would fly with a pair of Sidewinder armed Hawks. Once an incoming strike package had been detected by either Boeing E-3D Sentry AEW (Airborne Early Warning), a Royal Navy Type 42 AD (Air Defence) Destroyer, ground based radar or the Tornado's own Foxhunter radar the Tornado would set up an interception profile with its radar. If the

ZE161 FG from RAF Leeming based No.25(F) squadron taxis at RAF Leuchars top and deploys its thrust reverse buckets to aid braking bottom. Both Author

rules of engagement allowed, the Tornado F.3 would launch its Skyflash Missiles from BVR (Beyond Visual Range) in order to take out some of the enemy before the fighting got in close, after which it would close in on the enemy aircraft with the Hawks following. Even if the Skyflash missiles missed their targets, the target aircraft would become more vulnerable to short range infra red guided missiles while trying to evade the initial Skyflash shot.

In the MFF the Tornado F.3/Hawk T.1A combination proved a workable success, with the Hawk proving to be an extremely capable day fighter. Although the Hawk is slow, subsonic, compared with modern purpose designed fighters, it does have a number of advantages such as its

Top: ZE834 GM from No.43(F) squadron approaches for landing at Leuchars in 1996. Above: ZE858 GO from No.43(F) squadron taxis at Leuchars in 1993. Both Author

smoke free engine, small size and low radar signature, all combining to make the aircraft hard to detect both visually and with sensors such as radar. The Hawk has proved to be extremely agile with a manoeuvring limit of +9/-3g and while on

detachments to the Air Combat Manoeuvring Instrumentation (ACMI) range at Decimomannu, Sardinia, and the Hawk was used to winning its fight, whether it was against F-16 or F-15's.

During practice intercepts they worked closely with Ground Controlled Intercept personal (GCI) at RAF Buchan Sector Operations Centre (SOC) as well as RAF and NATO Boeing E-3 Sentry's. The information provided by Buchan or the E-3 allows

Top left: A Tornado F.3 from the OCU approaches to land in the early 1990's. **Centre top left: A No.43(F) squadron Tornado F.3 tucks up its undercarriage as it lifts-off from Leuchars in 1993.** Both Author **Bottom left and centre bottom left: Tornado F.3's take on fuel from RAF VC 10 in-flight refuelling tanker aircraft. Both** Crown Copyright **Top right: A No.111(F) squadron F.3 undergoes routine post-flight maintenance in a HAS at Leuchars.** Author **Above: A Tornado F.3 deploys flares for the camera.** G H Lee

Top: Flame and smoke erupts from the rear of a MBDA Skyflash BVR air to air missile as it ignites milliseconds after being pushed into the air stream by the still extended Frazer Nash rams on the belly of the Tornado F.3. BAE Systems **Above: A No.11 squadron Tornado F3 leaves a trail of smoke in its wake as it thunders down the runway during a Tactical Leadership Training exercise in 1995.** Author

A Tornado F.3 in Lima Fit sits in the QRA HAS at Leuchars in the 1990's. The aircraft is armed with pre-CSP load of four Skyflash and four Sidewinder air-to-air missiles. Above: A No.111(F) squadron

Tornado F.3 sits on the ramp at RAF Leuchars in the early 1990's. Both Author

the Tornado F.3 to reach a position from which a successful interception can be achieved. Such data will include target altitude and bearing relative to the friendly aircraft, as well as course instructions to enable them to reach a point where a radar contact with the on-board AI.24 Foxhunter can be achieved. Once radar contact has been confirmed they are on their own to conduct the combat. The Tornado F.3's practice interceptions against a wide verity of aircraft including Hawks, Jaguars, Tornado GR.1/4's and other Tornado F.3's.

The F.3 was second to none at low-level against a ground attack strike aircraft (commonly known as a 'mud mover'). Target aircraft locating the threat from the F.3 on their RHWR can often survive against Fox 1 (BVR) missiles launches. Therefore, a favoured F.3 tactic would be to launch a Fox 2 infrared guided air to air missile from within visual range as the target aircraft conducts a missile break manoeuvre to avoid the initial Fox 1.

During ACMI flights, Tornado F.3's often racked-up impressive results against modern fighters in the beyond visual range air to air arena. Kill rations of well over 20:1 against MiG-29 fighters have been achieved. However, in the close-range air combat arena the more agile MiG-29 and F-16 class fighters often gain the upper hand against the Tornado F.3. However, the F.3 is certainly no slouch, and can pull-off a few surprises. With wings in the 25-degree sweep position, with manoeuvre flaps deployed the F.3's turn performance is quite good, catching out many opponents during ACMI.

The Tornado F.3 was one of the first aircraft to be sent to Saudi Arabia following the Iraqi invasion of the Emirate of Kuwait on 2 August 1990. After sending a delegation to Saudi Arabia to persuade the Kingdom of a threat from Iraq and to allow coalition forces to be based on Saudi territory, the US began dispatching forces to the Gulf Region in early August 1990.

On 9 August, after reaching an agreement with the Saudi rulers, the UK government announced plans to send a force of aircraft to help deter Iraq from any advances against Saudi Arabia, although it is now clear that Iraq's campaign was not directed against Saudi Arabia. A force of Tornado F.3's was already in the Mediterranean as 11 aircraft from No.29 squadron were on armament practice camp at RAF Akrotiri, Cyprus. Six of these aircraft, ZE289 BA, ZE338 BB, ZE258 BE, ZE254 BG, ZE255 BH and ZE205 BF were flown out to Dharhan, Saudi Arabia on 10 August. A further 6, ZE762 CA, ZE758 CB, ZE163 CF,

ZE732 CH, ZE734 CJ and ZE736 CK from No.5 squadron joined these initial six aircraft on 12 August 1990.

Tornado F.3 ZE967 demonstrates the aircrafts low speed handling with air brakes deployed and undercarriage extended. RAF

A Tornado F.3 from Leeming based No.11 squadron banks into a starboard turn for the camera with wings in the full swept back position. The aircraft is configured with a full load of Skyflash and Sidewinder air to air missiles and a pair of 2250-litre external fuel tanks. Crown Copyright

Tornado F.3's from Coningsby based No.5 squadron taxi prior to a sortie. Crown Copyright

Above and below right: A RAF Tornado F.3 flies over the burning Kuwaiti oil wells following the cease-fire in March 1991. Both Crown Copyright

The Tornado's were based alongside No.29 squadron RSAF (Royal Saudi Air Force) also equipped with Tornado F.3s. Five of No.5 squadron's aircraft also joined the five remaining No.29 squadron Tornado's on Cyprus where they stayed until relieved by 6 Phantoms FGR.2's from No.19 squadron from RAF Wildenrath. BAe Rapier Surface to Air Missiles (SAM) from the RAF Regiment provided protection for the Tornado's and other aircraft based at Dhahran. However, as air defence cover at the base was sufficient, the Rapiers were re-directed to Muharaq Bahrain. Both half squadrons of Tornado F.3's were formed into No.5(C) Composite squadron, which worked closely with Tornado F.3's from the RSAF flying CAP's south of the Saudi, Kuwait, Iraq border.

As the unit had been dispatched at short notice, it was decided that No.5(C) squadron would be replaced by another unit taken from the Leuchars Wing, which had only recently received Tornado F.3's. The changeover began on 29 August, when eight aircraft, ZE961 DH, ZE203 DA, ZE208 DC, ZE210 DD, ZE936 DF, ZE962 DI, ZE968 DJ and ZE934 DV were flown out to Cyprus. Six of these

flew on to Dhahran the following day, allowing six No.5 squadron aircraft to return to the UK. The following week a further six modified Tornado F.3's were delivered to Dharhan allowing the remaining unmodified aircraft to return to the UK.

Most of the new Tornado's were from Block 13 production, ZE859 onwards. These aircraft had been delivered to the latest Stage 1 standard equipped with Stage 1 AA AI.24 Foxhunter radar. These aircraft had only just been delivered to the Leuchars wing from BAe Warton for the formation of No.111 squadron. The entire RAF operational F.3 fleet was eventually upgraded to at least Stage I standard by the early 1990's.

Tornado F.3 ZE966 from No.43(C) squadron is saluted by a Westland Wessex from resident B Flight No.22 squadron on its arrival back at Leuchars from the Gulf on 13 March 1991. The aircraft is carrying a full armament and a Phimat pod on the starboard outer stub station. Crown Copyright

For service in the Gulf region some Tornado F.3's were further upgraded to Stage 1+, which included the addition of improved cooling and revised software for the Foxhunter radar, including ECM (Electronic Counter Measures) and close combat capability. An F/A-18 type combat stick top, which contains all the buttons and switches needed for close combat engagements, was included in the Stage 1+ modifications. The aircraft's existing Marconi Hermes RHWR was also further improved. An important feature of the upgrade was the addition of an emergency power boost switch by the throttle, which could give an extra 5% power to each engine if required in an emergency. The standard RB199 MK104 was rated at 9,105-lb st dry and 16,523-lb st with afterburner. The modifications to the engine increased thrust by 5% by allowing the stator temperature to increase by 20 degree Centigrade from 1,290 degrees to 1,310 degrees Centigrade. The drawback of this higher thrust was a shorter time between overhauls.

The Tornado F.3 was modified to enable them to use Night Vision Goggles (NVG's), experience of which had been gained by the Tornado F.3 OEU. New tactics were developed for the use of NVG's, following which Tornado F.3 crews underwent a night work-up phase to familiarise themselves with their use. Another addition not normally worn in peacetime was the AP5 respirator, which is worn in flight protecting the crews from chemical attack. This may have been more a politically motivated addition as the main threat of chemical attack would be on the ground. Aircrew flying in pressurised cockpits at an altitude of 10-20,000-ft would be at little to no real risk in the air even if chemical weapons had been used. However, the respirator would, be required if the aircraft's base was attacked with chemical weapons and the F.3 could not divert to another base.

RAF Tornado F.3 ZE962 FJ in No.25 squadron markings taxis out of its shelter at Prince Sultan Air Base in Saudi Arabia. The aircraft has a full Skyflash and Sidewinder missile complement and a Phimat pod on the starboard outer pylon. Crown Copyright

The aircraft also received modifications enhancing their aircraft defence suite. A pair of Tracor AN/ALE40(V) flare dispensers was scabbed on underneath the rear fuselage on the engine-access doors for defence against infrared guided missiles. Each dispenser contains fifteen compartments. For protection against radar guided missiles, a Phillips-Matra (later MBDA) Phimat chaff pod was initially carried on the starboard wing pylon instead of an external fuel tank. The second external fuel tank was reinstated when the pod was relocated.

For operations in the hot climate the Tornado F.3's had their air conditioning systems uprated and the cockpit canopies were modified to prevent heat buckling, a problem that had plagued the original deployment. New tyres were introduced for use in the hot climate. Although the original Tornado deployment had used the standard AIM-9L a special purchase from the US of the improved AIM-9M for the Tornado F.3 improved the close combat capabilities still further. The AIM-9M has an improved seeker head, the WGU-4A/B and the NM36 mod II rocket motor.

The most visible of the changes was the addition of radar absorbent material (RAM). This was applied to the leading edges of the fin from the air scoop to the black dialectic panel at the tip, and the wing leading edges, apart from the dialectic panels for the RHWR located on the non-moving inboard part.

Ground crew check a RAF Tornado F.3 in No.11 squadron markings prior to a sortie from Prince Sultan Air Force Base in Saudi Arabia. Crown Copyright

On 14 September 1990, it was announced that a further six F3s, ZE203 DA, ZE907 DK, ZEI59 DO, ZE16S DU, ZE941 DW and ZE963 DX would be sent to Dhahran to reinforce the squadron already there. These arrived at the base on 22 September. On 1 December, the Tornado F3 detachment was taken over by personnel from No.43(F) Squadron Leuchars and No.29 squadron Coningsby, being

A RAF Tornado F.3 undergoes ground checks with a French Air Force Mirage F.1 reconnaissance aircraft in the background while based in Saudi Arabia for operations over the NFZ over southern Iraq. Crown Copyright

formed into No.43(C) composite Squadron. The squadron had been notified only on 29 November and within 48 hours was installed at Dharhan. On 2 December 1990, the new unit began flying CAP's in rotation with Saudi and USAF Tornado F.3s and F-15 Eagles. Between then and the 15 January 1991 United Nations deadline for Iraq to withdraw its forces from Kuwait or face 'all necessary means' from the coalition to remove them, the F.3 crews were busy flying CAPs and dissimilar air combat training sorties. Training sorties were flown against a number of aircraft types including French Mirage F.1's, Tornado GR.1's, Jaguars and US F/A-18 Hornets.

Just prior to the beginning of offensive operations on 17 January 1991, the F.3 squadron operation was moved to the protection of an underground bunker. At the same time the aircraft were given additional protection in the shape of splinter shields. As the hours ticked away to the beginning of Desert Storm, the squadron received the order to implement the offensive liberation of Kuwait, code word WOLFPACK, with H hour at 03:00 local on the morning of 17 January 1991. The Tornado F.3's were tasked with protecting Saudi territory from Iraqi bombers and with flying escort sorties for tankers and Boeing E-3 Sentry AWACS. For

this they were in standard war fit, armed with four under fuselage Skyflash BVR missiles, three AIM-9M Sidewinder short-range air to air missiles and a single 27-mm Mauser cannon. The fourth Sidewinder pylon was occupied by a Phimat chaff dispenser. The F.3's each carried two of the large 2250-litre external fuel tanks and had their already impressive CAP radius extended further by use of in-flight refuelling, normally from RAF Lockheed Tristar tankers.

A Tornado F.3 in No.11 squadron markings taxis at Prince Sultan Air Base, Saudi Arabia. For operations over the southern Iraq NFZ, RAF Tornado F.3's were armed with four Skyflash and four Sidewinder air to air missiles (later four ASRAAM missiles replaced the Sidewinders). For protection from surface to air radar guided missiles, a Phimat pod and an Ariel towed radar decoy were carried on the outboard wing stations. Crown Copyright

Above and below right: RAF Tornado F.3's operating over Bosnia Herzegovina during the civil war operated from temporary shelters at Gioa del Colle air base in Italy. The Tornado's are being prepared for sorties to enforce the NFZ over Bosnia. Both Crown Copyright

On 18 January, two F.3's on CAP committed north into Kuwait. They had been tasked to help a formation of USAF Fairchild A-10 Thunderbolt II ground attack aircraft which were escaping south, pursued by a pair of Iraqi air force fighters, which turned out to be Mirage F.1's. Once alerted the F.3s accelerated and blew off their external fuel tanks to reduce drag and increase launch energy for any possible missile launch. Once they had detected their targets the F.3s locked them up from long range with their powerful AI.24 Foxhunter radar. The Iraqis, after probably being alerted of the Tornado's presence by their RHWR, elected not to fight, instead turning north for safety.

On 18 February a pair of No.43 Composite Squadron Tornados' were vectored north by AWACS to intercept a pair of enemy aircraft that had been detected flying south fast at medium altitude. By the time the F.3s had closed to within 10-nm of the targets they had disappeared from AWACS radar. Although the F.3's still had radar

returns on their Foxhunter sets of the last known position of the contacts these looked more like chaff than actual aircraft. Moments later, another pair of aircraft were detected and the F.3's headed for them, but these turned out to be a pair of USAF F-4G Phantom II defence suppression aircraft. Both formations acknowledged each other's presence with the old fashioned wing rock and with the excitement over, the Tornado's headed back to Saudi territory over 150 miles distant. The return flight was far from boring as the RHWR, constantly alerted the F.3 crews that enemy radar was tracking them.

A pair of RAF Tornado F.3's escort a USAF Boeing (Rockwell) B-1B Lancer bomber as it takes on fuel from a USAF Boeing KC-10A Extender air to air refuelling tanker during a bombing mission over Iraq in March 2003. Crown Copyright

These incidents were two of only a few occasions that the F.3's confronted enemy aircraft. The main threat turned out to be from SAM and AAA once the F.3s patrols moved north. Crews were briefed to avoid the main SAM zones and flew above the ceiling of the smaller SAM's and most of the AAA. Just before the start of the ground war the F.3's CAP stations were pulled back behind coalition lines, allowing the artillery to fire at targets in Iraq without fear of hitting friendly aircraft on CAP.

Night sorties using NVG's were a common practice for the F.3 detachment, requiring great skill from the crews, as aircraft had to fly with lights out for obvious reasons. It was during a night sortie that a pair of Tornado's flying CAP on the Saudi Kuwait Iraq border was ordered to proceed some 60 miles into Iraq. As there was no mention of enemy aircraft the F.3 crews enquired as to why. They were told that they were to assist in the search for a downed allied aircraft, as their NVGs would be of considerable use. AWACS had passed the coded frequency to the F.3's coded 'Bentley' in

order for them to try and contact the downed aircraft. This was to no avail, therefore, Bentley 01 dropped down to low level in the hope of locating them visually while Bentley 02 remained at medium altitude to provide cover against any Iraqi air threat. After some 40 minutes on station lack of fuel dictated that Bentley 01 and 02 depart, leaving the search operation in the hands of AWACS and other allied aircraft in the vicinity.

As the war progressed CAP's were flown over Kuwait and Iraq. With the cease-fire signed on 3 March, F.3's flew CAP's until stood-down on 8 March, returning to the UK on 13 March 1991. While no air to air kills were achieved, this did not reflect on the squadron's contribution to the campaign. The Iraq air force at a technological and numerical disadvantage opted not to fight on mass. Most of the USAF kills were against aircraft fleeing to Iran. The only coalition air to air loss was a US F/A-18 fighter shot down by an Iraqi fighter on the first day of the war.

Since the 1991 Gulf War ended, there has been much written about the lack of air to air kills achieved by the Tornado F.3 detachment. Many excuses have been put forward about how the Tornado F.3 was held back so that USAF F-15 Eagles could gain kills. These suggestions like similar suggestion concerning the USN F-14

RAF Tornado F.3's take on fuel from a RAF Lockheed Tristar air to air refuelling tanker on 20 March 2003; the first day of the invasion of Iraq. Crown Copyright

Tomcats are simply ill founded and ill researched. The Tornado F.3 deployment conducted exactly the type of mission it was designed for, procured for and deployed to the Persian Gulf for. That is combat air patrols to protect friendly territory and friendly ground forces from air attack. The USN F-14 Tomcats role in the 1991 was air defence of the fleet and tactical reconnaissance, exactly the type of mission they conducted. The USAF F-15A/C Eagles which achieved most of the air to air kills against Iraqi fighters were designed primarily as counter air fighters designed to achieve air superiority over the battlefield or enemy territory. This is the type of mission that was conducted by the F-15 during the war. Therefore, this volume will not make excuses for the lack of shoot downs conducted by the Tornado's as no excuse is necessary for a detachment which successfully conducted the mission for which their equipment was designed and their deployment called for. Contrary to some comments that the RAF Tornado F.3 detachment only flew defensive CAP's south of the Saudi Arabian border, as noted in the text above, RAF Tornado F.3's did actually fly a number of sorties over Iraq and Kuwait during the 1991 war.

RAF Tornado F.3 units conducted a number of deployments to the Gulf Region to support US and British operations over the controversial NFZ (No Fly Zones) over Iraq, which were initially set up by the US, Britain and France without United Nations authorisation. There was a number of bombing operations flown against Iraq during the early to late 1990's. The RAF's main contribution to these operations known under the British code names of operation Warden for the northern NFZ above the 36th parallel and operation Bolton and Resonate south for the southern NFZ below the 32nd and later the 33rd parallels. From 31 December 1996, France withdrew from the northern NFZ operation as it stated that the operation was no longer being conducted to protect Kurds and had instead simply become an operation of surveillance of Iraq. French aircraft remained based in the Persian Gulf. From December 1998, the NFZ operation effectively entered into a low key bombing campaign, which lasted over three years and achieved extremely little in terms of damage inflicted upon the Iraqi military. The operation, however, did achieve much in the way of international condemnation particularly within the ranks of the United Nations, many of whose member nations were outraged at the civilian targets, which were bombed and civilian casualties, which resulted from these bombing missions.

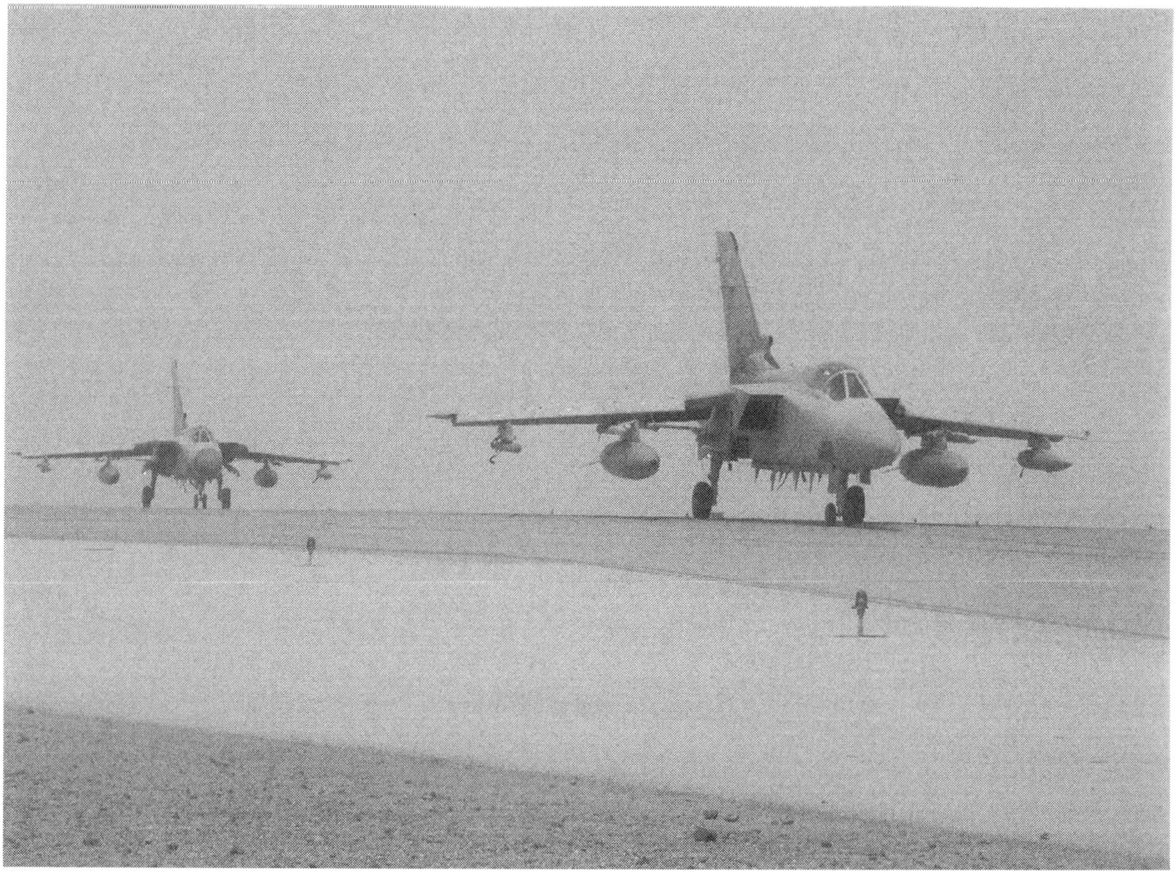

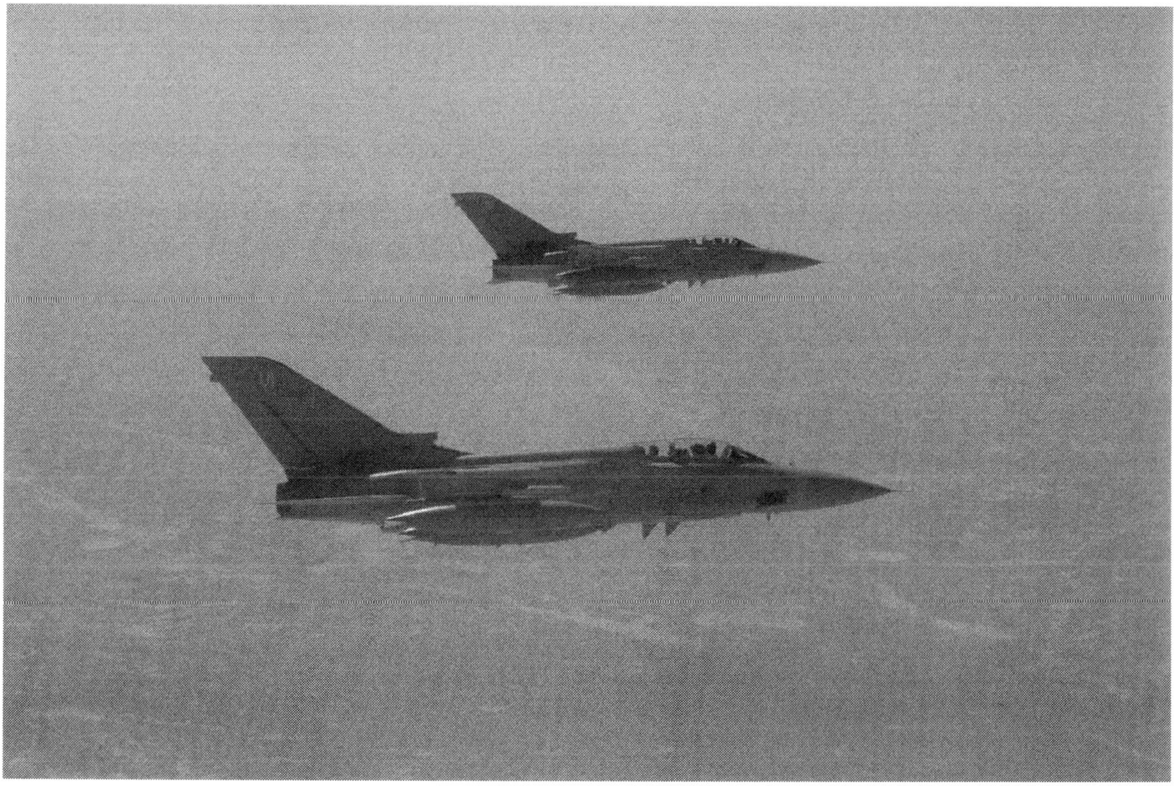

Previous page and this page: RAF Tornado F.3's were employed in Operation Telic, the British contribution to the United States, United Kingdom, and Australian invasion and occupation of Iraq in 2003. RAF

.

A RAF Tornado F.3 is silhouetted against the sun during a sortie in support of the March 2003 invasion of Iraq. Crown Copyright

During this low-key bombing campaign, Saudi Arabia refused to allow British or US aircraft bombing Iraq to operate from its territory. However, RAF Tornado F.3's was allowed to operate from Saudi Arabia, as their mission was considered defensive in nature and not to drop ordnance on Iraq.

The RAF contributed a force of Tornado F.3's to the US; British and Australian force which invaded and occupied Iraq without United Nations authorisation causing worldwide condemnation for the unprovoked attack and invasion of a sovereign state. The RAF campaign under the code-name Operation Telic commenced on 20 March 2003, but with the Iraqi Air Force hardly capable of being considered a viable force, equipped with outdated technology and obsolete aircraft, the Tornado F.3's like other coalition air defence assets were not challenged in the air. The F.3's settled down to a short campaign of escorting strike and support aircraft participating in the operation.

RAF Tornado F.3's contributed to the NATO force enforcing the Air Exclusion Zone (AEZ) over Bosnia Herzogovina during the Bosnian civil war of 1992-1995. Six Tornado F.3's were sent to Gioa del Colle air base in Italy in April 1993. The deployed F.3's were to a similar Stage 1+ standard to those deployed to the Persian Gulf in 1990/91. One of the few differences was the weapons load carried. In the Gulf, the Tornado F.3's carried three AIM-9M's and a single Phimat chaff dispenser pod on the inboard wing stub-pylons. Aircraft deployed to enforce the Bosnian AEZ carried two AIM-9M's and two Phimat chaff pods. The F.3's also carried the standard load of four Skyflash BVR missiles.

For patrols over Bosnia, two CAP stations were established. The first CAP station was located in the north of the country covering Banja Luka/Tuzla and the surrounding areas. The other CAP station over the south of the country covered the Mostar Sarajavo areas. In-flight refuelling was required to keep the fighters on station in the CAP areas. To support the fighters a 24-hour tanker track was established near the Croatian City of Split.

A RAF Tornado F.3 takes on fuel from an USAF KC-10 Extender tanker aircraft as F/A-18 Hornets wait in the background during a mission in support of the invasion of Iraq on 20 March 2003. Crown Copyright

The first Tornado F.3 sorties were flown on 26 April, and the first attempted intercept of a violating aircraft took place on 29 April when a pair of Tornado F.3's was vectored towards a contact, which was lost before the fighters could visually identify it.

On 13 May 1993, two F.3's were sent to investigate a Croatian Mil Mi-8 Hip transport helicopter, which was subsequently forced to land. The aircraft, supposedly on a casualty evacuation mission, was found to be carrying 20,000 rounds of ammunition. The dangers of this type of operation were that if the mission had been a genuine casualty evacuation mission, forcing it to land before it reached its objective could put the casualty's life at risk.

Throughout 1993, the three squadrons of the Leeming wing contributed aircraft and crews to the detachment, which was then taken over by the Coningsby wing on 25 November 1993, starting with No.5 Squadron, followed by No.29 Squadron in March 1994. The two squadrons of the Leuchars wing then took over the detachment starting with No.43 Squadron in May 1994 followed by No.111

Squadron before the cycle of unit rotations began again.

The Bosnian conflict was more or less ended with the Dayton peace agreement signed in late 1995. This followed a chain of events, which began in summer 1995 leading to NATO launching a limited air offensive against Bosnian Serb targets under the code name operation 'Deliberate Force'. While this is often credited with forcing a negotiated end to the conflict, in reality the air offensive contributed only a small part to the pressure Bosnian Serb forces were put under in late summer 1995. A powerful Anglo-French Rapid Reaction ground forces complete with artillery and heavy armour put pressure on Serb forces to break the siege of Sarajeavo. The artillery of the Anglo-French force was in action against targets even on days when NATO aircraft did not strike any targets. This offensive combined with a Croatian offensive under the Code Name 'Operation Storm' severely weakened the Bosnian Serb forces paving the way for the Dayton peace agreement.

With Bosnian cease-fire in place, NATO began to wind down its operations over the country. The Tornado F.3 detachment was withdrawn back to the UK on 5 March 1996.

Britains air defence-team during the 1990's and well into the first decade of this century! A Tornado F.3 from the Tornado F.3 Operational Evaluation Unit flies in formation with a RAF Boeing E-3D Sentry AEW MK1. Crown Copyright

Once deliveries of the Tornado F.3 were well underway, the Tornado F.2's were withdrawn from service with No.229 OCU. The first Tornado F.2 went to RAF St Athan for storage in March 1986, with the last of the initial aircraft allocated for storage following in January 1988. Initial plans to upgrade the Tornado F.2's to F.2A standard, which would have been identical to the F.3 except for the F.2 retaining the RB199 MK103 engines, were abandoned with the decision for the RAF to take delivery of eight Tornado F MK3's originally ordered for Oman.

In summer 1992, it was being speculated that the Tornado F.2's in storage at RAF St Athan could be leased to the Royal Malaysian Air Force (RMAF) pending a future sale of Eurofighter (then known as EFA). However, Malaysia's requirement was for around 24 aircraft and with only 15 Tornado F.2's in storage such a deal seemed unlikely. Malaysia went on to purchase eight McDonnell Douglas F/A-18D Hornet strike fighters and 18 RSK MiG-28N's for air defence. In 1993, it was revealed that 12 of the Tornado F.2's in storage; ZD901, ZD903, ZD904, ZD905, ZD932, ZD933, ZD934, ZD936, ZD937, ZD938, ZD940 and ZD941 would be scrapped

In 1993, the RAF had to ground eighteen of its Tornado F.3 fleet after the discovery of some serious damage to the Airedales following servicing by a private contractor. The MoD had contracted out airframe modification work to increase the Fatigue Index (FI) of its Tornado F3 fleet. The first contract for 15 aircraft, which was won by BAe, had already been carried out satisfactorily. The second such contract for a total of 18 aircraft was won by Airwork Services for a price of £7 million, which was £4 million less than BAe. With the discovery of the damage the MoD immediately cancelled the contract.

Tornado F.3, ZE839 from No.56 Reserve Squadron is seen at RAF Leuchars in September 2004. No.56 (Reserve) Squadron had moved to the Scottish base the previous year Author

A report on the work, which was being conducted at RAF St Athan stated that due to delays in completing the contract Airwork personnel began using inappropriate tools for the work. Apparently longerons were distorted when pneumatic guns similar to those used to remove car wheels were used to remove light alloy collars covering the nuts and bolts holding panels to the longerons, causing the collars to be almost chiselled off. RAF

The first production Tornado F.3, ZE154 seen getting airborne in full afterburner just after re-delivery in 1995, was the first of the Tornado F.3's to be rebuilt by BAe at Warton using the centre fuselage sections of retired Tornado F.2's. ZE154 received the centre fuselage section of Tornado F.2 ZD901, which was fitted with the damaged centre section of ZE154. Following successful test flights, the aircraft was delivered to No.43(F) squadron at Leuchars in late summer 1995. Author

maintenance personnel discovered the damage after pilots reported problems with the first four aircraft returned to operational service. This resulted in tests being done on the remaining fourteen aircraft at RAF St Athan where twelve were found to be seriously damaged while the other two were slightly damaged. In late May 1993, officials from DASA (now EADS Germany) inspected the aircraft and were horrified at what they saw.

The four aircraft that were delivered to their units ZE292, ZE295, ZE343 and ZE728 were subsequently returned to St Athan for repair. The 25 FI structural updating program was immediately halted after the discovery of the damage but RAF personnel had the conversion line going again four weeks later and in the first half of 1994 three aircraft ZE163, ZE200 and ZE789 had been returned to service with No.56R, No.11 and No.56R, squadrons respectively. On 16 September 1994, the RAF was awarded the follow-on contract for the updating program.

By early 1994, it was being reported that 14 of the 18 aircraft affected were to be prematurely scrapped as it was suggested that the 12 most damaged aircraft might have had to be sent to Germany in order for their centre fuselages to be rebuilt in the original manufacturing jigs to restore structural strength. Another less costly alternative was to use the centre fuselages from the Tornado F.2's stored at RAF St. Athan. Therefore, the decision to scrap the F.2's was cancelled in favour of utilising the centre sections from the retired aircraft to repair the damaged F.3's.

Ground crew drag a fuel pipe from the ground tanker to refuel a Tornado F.3 in No.11 squadron markings as the aircraft is prepared for a sortie at a forward-deployed location. Crown Copyright

ZE966 DZ from Leeming based No.11 squadron banks left during a training flight over the North Sea in the early 1990's. Crown Copyright

During late 1994, BAe started repair work on the first of the more seriously damaged of the 18 aircraft damaged by Airwork in 1993. Three less seriously damaged aircraft mentioned above having already been repaired and returned to service. BAe was given a contract for a trial rebuild of Tornado F.3 ZE154 using the centre section of Tornado F.2 ZD901. Both these aircraft were in storage at St Athan from where they were taken by road to BAe Warton on 24 October 1994. Of the 18 F.2s built (excluding the three prototypes) ZD899 and ZD939 were serving with the MOD (PE) BAe Warton, ZD900 was with the A&AEE Bascombe Down and ZD902 was with the DRA also at Bascombe Down. This left the other 14 aircraft, 13 of which were stored at RAF St Athan and the other at RAF Coningsby, ZD901, ZD903 - ZD906, ZD932 - ZD938 and ZD940 - ZD941 available for use in the rebuild program. ZD900 and ZD939 were retired allowing them to be used in the rebuild program.

The first rebuild trial was successful with ZE154 and fuselage centre section of ZD901 conducting test flights at BAe Warton before being re-delivered to the RAF for service with No.43(F) squadron at Leuchars in the late summer of 1995. The second aircraft to be repaired was ZE295 using the fuselage centre section of ZD938, followed by ZE254 using the fuselage centre section of ZD941. The other F.3/F.2 tie-ups are as follows: ZE294/ZD906, ZE255/ZD932, ZE251/ZD936, ZE786/ZD934, ZE258/ZD905, ZE288/ZD940, ZE292/ZD939, ZE343/ZD900, ZE728/ZD903, ZE729/ZD933, ZE736/ZD937, ZE759/ZD904 and ZE793/ZD935. The Tornado F.2s received the damaged F.3 fuselage centre sections and were returned to St Athan. The last of the rebuilt aircraft was delivered to the RAF in 1997.

Known fates of the three ADV prototypes and 18 Tornado F.2 production aircraft include the following. The first Tornado ADV prototype A01, ZA254 was transferred to RAF Coningsby where the fuselage was used for ground instruction training by late 1996. The second prototype A02, ZA267 was sent to RAF Marham for use as a

No.111(F) squadron Tornado F.3 ZE159 UV flies with a load of four MBDA ASRAAM air-to-air missiles, which entered operational service with the RAF in 2002. G H Lee

ground instruction trainer. The third Tornado ADV prototype, A03, ZA283 was with the DPA/AFD at Bascombe Down in late 1999 early 2000, but was put into storage and eventually scrapped at RAF St Athan in January 2001. Tornado F.2 ZD900 was with the Leuchars BDRT by late 2000; the fuselage was used for ground instruction at RAF Lossimouth in summer 2001, and was with a private owner in chesterfield for scrapping in early 2003. The fuselage was also possibly used by St Athan fire section for a time. Tornado F.2's ZD904, ZD937 and ZD941 were scrapped at St Athan in January 2001. ZD906 was also reported scrapped in January 2001, however, the aircraft was also reported on the strength of RAF BDRT as late as 2003. The fuselage of ZD905 was allocated for scrapping with a private owner by early 2003. The fuselage of ZD933 was at St Athan for ground instruction by early 2002. The fuselage of ZD934 was stored at St Athan by late 2000, but was transferred to RAF Leeming as a ground instruction airframe in 2001. ZD935 was stored at

RAF Shawbury then transferred to St Athan for use by the fire section in 2000. Tornado F.2 ZD936 went to EADS Germany in 2001, but was displayed at the Bascombe Down Museum by late 2003. ZD938 was with the St Athan BDRT by late 1998 and was reported stored at Shawbury by summer 2000. ZD939 was moved to RAF Cosford as a ground instruction airframe by early 2002. The fuselage of ZD940 was with St Athan fire section by early 2003.

A No.56(R) Squadron Tornado F.3 about to be refuelled at RAF Leuchars in September 2004. Author

Top and above; No 56(R) Squadron moved from Coningsby to Leuchars in late March 2003 and in 2008 it merged with No.43 Squadron. Both Author

In with the new and out with the old! Not quite yet, however, Eurofighter Typhoon development aircraft DA2 (ZH588) sits alongside a Tornado F.3 during a deployment for development flights at Leuchars in 2001. The Typhoon has began entering RAF service and should have completely replaced the Tornado F.3 in the air defence role by 2010, when the last Tornado F.3's are scheduled to be withdrawn from operational service. Eurofighter GmbH

With the disbanding of three of the seven operational front line squadrons to operate the Tornado F.3 and the return of the survivors of the 24 aircraft leased to Italy, the RAF has also reduced the number of F.3's in service. A number of aircraft have been lost in accidents including one of the aircraft leased to the Italian Air Force. Among Tornado F.3's withdrawn from operational flying service are ZE210, which was used as a spare source at Leuchars before the fuselage was scrapped at St Athan in January 2001. ZE758 (ZE343) - ZE343 was marked as ZE758 by September 2001 - was allocated the maintenance serial 9298M and transferred to No.1 School of Technical Training (SoTT) at Cosford by late 2001 for ground training. ZE760 was returned to the RAF by Italy by late 2003 and allocated for display at Coningsby. ZE967 was installed as a gate guard at RAF Leuchars on 9 September 2004.

The first Tornado F.3 squadron to disband was Leeming based No.23 squadron. On 7 July 1993,

the MoD announced that No.23 squadron was to disband the following year, with the squadron officially disbanding with the Tornado F.3 on 23 March 1994. This left the RAF with only six front line fighter squadrons along with the F.3 OCU. In 1996, No.23 squadron re-formed as part of the Waddington based Boeing E-3D Sentry AEW MK1 force sharing its aircraft with No.8 squadron, which was already operating the aircraft. The next Tornado F.3 unit to go was Coningsby based No.29 squadron, which was disbanded on 30 October 1998. No.29 squadron re-formed as a Eurofighter Typhoon squadron at RAF Coningsby in 2003. No.5 squadron also based at Coningsby was disbanded in 2002. In July 2004, defence cuts announced by the MoD included the early retirement of one of the four remaining Tornado F.3 squadrons. As the two operational squadrons at Leuchars are tasked with the only permanent 24-hour a day, 365 day a year QRA in the UK, it was inevitable that the Tornado F.3 squadron to disband would come from the Leeming wing AND No.11 Squadron was disbanded in October 2005, reforming on 29 March 2007 with Eurofighter Typhoon F.2's. The remaining three squadrons and the OCU would continue to serve the RAF with units being withdrawn as Eurofighter Typhoon began to come on line in strength. The last Tornado F.3 squadron would serve until the type was retired in 2011.

Chapter Four

Tornado ADV Exports

The Royal Saudi Air Force was the first and only non-Panavia partner nation to order and receive either the IDS or ADV variants of the Tornado. The Kingdom eventually ordered and received a total of 96 Tornado IDS and 24 Tornado ADV. One of the 24 Tornado ADV's is seen during a test flight prior to delivery to the RSAF. BAE Systems

Although not the first export customer to order the Panavia Tornado the *I Quwwat al Jawwiya as Saudiya* (Royal Saudi Air Force-RSAF) was the first export customer to order both the IDS and ADV variants of the Tornado, as well as being the first to receive export Tornado's. The RSAF had a tradition of buying British military aircraft, having previously received BAC Strikemaster ground attack aircraft and BAC Lightning fighters. On 26

September 1985, the Saudi Government signed an order for 48 Tornado IDS, equivalent to the RAF's GR MK1, and 24 Tornado ADV, equivalent to the RAF's F MK3. The *Al Yammannah I* contract also included a pair of BAe Jetstream twin turboprop trainers fitted out as navigator trainers for both variants of the Tornado as well as 30 each of Pilatus PC-9 turboprop trainers and BAe Hawk MK 65 advanced trainers. The agreement also called for early delivery of 20 aircraft. For this reason the RAF had to give up 18 of its final GR MK1s, with the other two aircraft originally being manufactured for the then West Germany. All these aircraft were originally in Batch 7, but were brought forward to be included in Batch 6.

The first Tornado F MK3 to be delivered to the RSAF was 2906 (MoD serial ZE859), which conducted its first flight on 1 December 1998, with delivery to the RSAF taking place in February 1989. The aircraft is seen here carrying a load of four Skyflash and four Sidewinder air-to-air missiles. BAE Systems

The 24 Tornado ADV's for the RSAF were completed from the RAF's order included in Batch 6. All of these aircraft were single stick aircraft, although six were completed as twin-stick operational trainer aircraft to meet the RSAF order for six such aircraft, as the RAF was unwilling to part with any twin-stick aircraft. The RAF received ADV's from later batches to make up for the loss of its 24 aircraft. The first Tornado F.3 for the RSAF was ZE859 with the Saudi serial 2906. This aircraft flew for the first time on 1 December 1988, and was subsequently delivered to Saudi Arabia on 9 February 1988. The aircraft was handed over to the RSAF to equip No.29 squadron RSAF based at Dahrhan on 20 March 1989, along with three other F.3's. No.29 squadron RSAF had received its full complement of 12 F.3's, including four dual control aircraft by 20 September 1989. The second Tornado ADV squadron for the RSAF was No.34 squadron, which was also based at

Dhahran. This unit received its first aircraft on 14-15 November 1989, although it was initially assigned to co-located No.29 squadron. No.34 squadron completed its re-equipment with the Tornado ADV in 1990.

RSAF Tornado F.3's were the first to be fitted on the production line with the Stage One version of the AI.24 Foxhunter radar having the specification called for by the RAF. In squadron service Saudi Tornado F.3's were essentially equipped the same as their RAF cousins with the primary armament of four BVR Beyond Visual Range) Skyflash air-to-air missiles complemented by four short-range infrared guided AIM-9L Sidewinder air to air missiles. Saudi aircraft were also equipped with a single 27-mm Mauser cannon. The RSAF order also included supply of the large 2250-litre (495 imperial gallons) external fuel tanks, which could be substituted with the 330-gallon tanks used by the Tornado IDS fleet. The RSAF adopted a similar overall air defence grey colour scheme to the RAF F.3's. Markings are carried in the shape of Saudi roundels on both sides of the forward fuselage and on the upper port wing and lower starboard wing USAF style. The starboard upper wing has the letters RSAF with the same on the port lower wing.

A RSAF Tornado F.3 reveals its underside showing its primary armament of four Skyflash BVR missiles. The Tornado F.3's delivered to the RSAF were among the most advanced combat aircraft delivered to the Persian Gulf Region at the time of their delivery. BAE Systems

RSAF Tornado F.3 interceptors share their role with the Boeing (formerly McDonnell Douglas) F-15C/D Eagle fighters delivered from the US. The Tornado F.3's long-range and endurance gave the RSAF an extremely capable air defence aircraft capable of performing long duration CAP's or projecting an air defence capability at long range in the Gulf Region.

Another Tornado sale to the RSAF under *Al Yammannah II* was provisionally agreed in July 1988, but was not formally signed until May 1993. This second Tornado order was originally for 12 Tornado IDS and 36 Tornado ADV variants, but when signed it was altered to consist of 48 Tornado IDS along with a further batch of Hawk trainers.

At the outbreak of hostilities between Iraq and Kuwait on 2 August 1990, the RSAF began mounting defensive CAP's south of the Saudi border with Kuwait and Iraq as it was unknown whether Iraq had further ambitions in the region. Sharing in these duties were the McDonnell Douglas F-15C/D Eagles of the RSAF No.13 and No.42 squadrons. The Saudi Tornado F.3 and F-15's were soon joined by American and British interceptors within a week or so. During Desert Shield, the build-up of coalition forces in the Persian Gulf the Saudi Tornado F.3's worked with their RAF counterparts, both undertaking daily CAP's along with other Coalition fighters.

A RSAF Tornado IDS is seen on a test flight before delivery from BAE. The IDS was the main Tornado variant operated by the RSAF, 96 of-which were delivered under the Al Yammannah I and II contracts. BAE Systems

Previous page: RSAF Tornado F.3's participated in the 1990 Operation Desert Shield and 1991 Operation Desert Storm alongside RAF Tornado F.3's and RSAF and USAF F-15 Eagles. All US DoD **Above: RSAF Tornado F.3 2906 during a test flight prior to delivery.** BAE Systems

When war arrived in January 1991, the Tornado F.3's worked long hours in both defensive and offensive CAP's with the crews having no resulting air-to-air combats to show for their work. This was due to the fact that the Iraqi air force was technologically outclassed by and unable to effectively confront the modern air power arrayed against it.

The first customer for the Tornado ADV was the *Al Quwwat al jawwiya al Sultanat Oman* (Royal Air Force of Oman), which ordered eight aircraft for the interception role in order to release the service's SEPECAT Jaguars - which were then employed on interception duties - for the ground attack role. The aircraft were to be dual control under an order placed on 14 August 1985. Deliveries were to have begun in 1988, but this date was put back to 1992, until the order was

finally cancelled with the RAF taking delivery of the eight aircraft instead. The aircraft received the RAF serials ZH552-ZH559, with the last being delivered in March 1993.

RSAF Panavia Tornado ADV serial Numbers	
2901 (ZE861)	2902 (ZE881)
2903 (ZE882)	2904 (ZE883)
2905 (ZE884)	2906 (ZE859)
2907 (ZE860)	2908 (ZE885)
2909 (ZE886)	2910 (ZE890)
2911 (ZE891)	2912, (ZE905)
3451 (ZE909)	3452 (ZE910)
3453 (ZE912)	3454 (ZE913)
3455 (ZE914)	3456 (ZE935)
3457 (ZE937)	3458 (ZE938)
3459 (ZE939)	3460 (ZE940)
3461 (ZE943)	3462 (ZE960)

2901-2904 and 3451-3452 were twin stick dual control aircraft, although still retaining full operational capability like their RAF counterparts

The first Tornado F.3 for the Italian Air Force, MM7202 flies in formation with No.56(Reserve) squadron Tornado F.3 ZE839 AR. RAF

By 1987, BAe was trying hard to interest the JASDF (Japan Air Self Defence Force) in the Panavia Tornado J. This would have been jointly developed with the Japanese aircraft industry. The Tornado J was developed from the earlier Super Tornado, which would use the ADV's fuselage, but with IDS and ECR equipment included. If successful the Tornado J could have been used for long range maritime attack with the Mitsubishi ASM-1 anti-ship missile, but would have retained full interception capability. To demonstrate the feasibility of such an aircraft the third ADV prototype ZA283 had its outboard wing stations activated, and flew a series of demonstration flights. However, the Tornado J came to nothing with the JASDF eventually selecting a variant of the General Dynamics (now Lockheed Martin) F-16 Fighting Falcon, which was developed into the Mitsubishi F-2.

In June 1993, the Aeronautica Militari (AMI/Italian air force) announced that it would like to purchase or lease existing fighters from some of its NATO allies in order to counter the potential air threat from Serbia. For this the Italian Air Force then relied on obsolescent Lockheed F-104S Starfighters, which were at that time scheduled for replacement by Eurofighters at the end of the century. As Eurofighter was still 7-10 years from

entering service the Italians needed an interim fighter to plug the air defence capability gap made more urgent due to the conflicts in the Balkans. Front runners were surplus USAF F-15 and F-16 fighters and RAF Tornado F.3's made surplus by defence cuts. On the one hand, it would have been more realistic for the Italian air force to choose the US fighters as they were single crew aircraft like the F-104 and in the case of the F-16, single engine. On the other hand, the Italians already operated the Tornado IDS and ECR (Electronic Combat and Reconnaissance) variants in the strike and reconnaissance roles and, therefore, had air and ground crews with considerable Tornado experience and infrastructure to support the new Tornado variant.

The Italian requirement was for up to 50 aircraft and the USAF offered up to 70 F-15 and F-16s for lease or sale. The RAF on the other hand would have been severely over stretched to maintain the then proposed six squadrons and an OCU as well as an out of area flight in the Falklands if it had had to part with 50 of its Tornado F.3 fleet

On 6 November 1993, the Italian Defence Minister, while on a visit to London, announced that the Tornado F.3 had been selected as the Italian Air Forces stopgap fighter pending the introduction of Eurofighter. The UK MoD had held discussions with their Italian counterparts in connection with the lease of 24 (not the 50 previously mentioned) Tornado F.3s over a ten year period. In March 1994, an MoA

36° Stormo was the first Italian air force unit to be equipped with the Tornado F.3. The unit received its first aircraft on 5 July 1995. Author

(Memorandum of Agreement) was signed covering the lease of 24 Tornado F.3's including four twin-stick aircraft, 48 RB199 MK104 engines and 96 Skyflash missiles, which were apparently to a reduced standard compared with RAF Skyflash stocks. The contract also covered training, maintenance facilities at RAF St Athan and logistics supply. The lease deal was flexible allowing an extension, early termination or even Italian purchase of the aircraft. The lease of the Skyflash missiles as the Italian Tornado F.3 main armament put to rest rumours that the aircraft would be modified to fire the Alenia (now MBDA) Aspide BVR semi-active radar homing air to air missile, which was used on the Italian Air Force Lockheed F-104 ASA(M). The leased aircraft were all at least Stage One+ standard and had gone through the RAF's 35FI (Fatigue Index) structural enhancement program. All 24 aircraft still had over half of their design 4,000 flight hour airframe life remaining with typical hours flown being in the region of 1500 hours. The lease deal also covered further planned improvements to the aircraft

including the introduction of JTIDS and the Stage 2-radar upgrade. JTIDS was introduced in late 2000. Among the few changes made to the Italian aircraft was the replacement of the RAF's electronic warfare and RHWR library with Italian equipment.

By March 1994, it was decided that 12o Groupo (36o Stormo at Gioia-del-Colle) and 1o Groupo (37o Stormo at Trapani/Birgi) would be the units to re-equip with Tornado F.3s. Unusually it was also announced that the aircraft would be flown by two-pilots who would alternate between the front and rear cockpits. The 24 aircraft were to be delivered in two equal batches with the first aircraft ZE832 (Italian serial MM7202) with side number 36-12 being delivered to Gioia-Del-Colle air base in Italy on 5 July 1995. The last aircraft of the first batch of 12 aircraft was delivered in January 1996 completing re-equipment of the 36o Stormo, which was declared operational in June 1996. The second batch of 12 aircraft was delivered to Italy between February and July 1997 and the unit was declared operational in 1998.

Italian Tornado F.3's were assigned to the National Alert Service for air defence of Italy. With the crisis situation in the Balkans throughout the 1990's, this task took on greater urgency for

Italy in part prompting the lease of the new fighters. The quick reaction alert was conducted from Gioia-del-Colle where 36o were based. 21o were deployed to Gioia on regular long-duration deployments to participate in the QRA duties. In March 1999, 21o Gruppo was allocated autonomous status and moved to Gioia del Colle joining 36o Gruppi concentrating Italian Tornado F.3's at one base. With 21o Storm having already moved to Gioia del Colle, 53o Stormo was disbanded in July 1999, with 21o Gruppo coming under full control of 36o Stormo

Top: Italian Tornado F.3's normally had no tail markings as seen on the tail of MM7211. In 2002, MM7234 was pained in a special Tiger colour scheme, which included a large black and white, striped Tiger on the tail with a blue outline. Both Author

Maintenance problems had plagued the Italian Tornado F.3 fleet since delivery and the fact that Italy did not lease any spare engines exacerbated the problem further. When operation 'Allied Force', (the NATO bombing campaign against Serbia, Kosovo and Montenegro) began in March 1999, the RAF delivered four spare RB199 MK104 engines to Gioia-del-Colle to help improve the Italian's poor serviceably rate for the Tornado F.3. During the March-June 1999 campaign, Italy

assigned a maximum of 6 Tornado F.3's for air defence at any one time. The Italian F.3's main role was CAP to protect Italy from any retaliatory attack from Serbia, however, unlikely this was due to the fact that Serbia lacked the resources for such an endeavour. With the Serb air force in such a poor state of serviceability and equipped with obsolescent equipment, the Italian Tornado F.3's were never challenged during CAP's. Rumours emerged that on 9 April 1999, a pair of Serb MiG-29 fighters challenged a pair of Italian F.3's on CAP, although the Italian government denied this. In the absence of any evidence to the contrary, this volume will conclude that such an encounter did not take place. The four loaned RAF RB199 MK104 engines were returned to the UK later with the AMI bringing four RB199 MK103 engines up to more or less MK104 standard to replace them.

In March 2001, 21o Gruppo was disbanded and officially merged with 12o Gruppo, which took over 21o Gruppo's Tornado F.3's.

Italian Tornado F.3's wore a roundel on the forward fuselage just between the two cockpits. Author

Further delays to the full introduction to squadron service of the Eurofighter Typhoon and the increasing obsolesce of the AMI's F-104 fleet posed serious problems for the AMI. The Starfighters were in desperate need of retirement and the Tornado F.3's would require modernisation, particularly to their weapon capability if they were to remain viable air defence platforms throughout the first decade of the 21st Century. With the decision to retire the F-104, the AMI would require additional stopgap fighters pending deliveries of Eurofighter Typhoon. The options appeared to be to put the Tornado F.3's through the RAF's CSP (Capability Sustainment Program) providing the Raytheon AIM-120 AMRAAM active-radar guided air-to-air missile

A pair of Italian Air Force Tornado F.3's on CAP refuel during operation Allied Force in March 1999. USAF

or purchase or lease a new interim fighter. If the Tornado option had been pursued, then the AMI would have required additional fighters, which the RAF would have been hard pressed to provide. The unwillingness to fund the CSP for the Tornado F.3 led to the announcement that the AMI would lease 34 Lockheed Martin F-16A/B fighters, which had been retired by the United States. The decision was taken to terminate the Tornado F.3 lease early with aircraft returning to the UK from early 2003 and continuing into late 2004.

That the Tornado ADV was not an outstanding export success story was no fault of the aircraft itself, but rather a combination of politics, high price (flyaway price about £22 million at late 1980's early 1990's prices) and the complexity of the design. Most Western influenced nations went instead for cheaper, simpler and less capable designs such as the US Lockheed Martin F-16 Fighting Falcon. Indeed, with the F-15 Eagle, F-16

and the F/A-18 Hornet, the US manufactures had the Western influenced fighter market cornered from the early 1980's, with such large home production runs allowing lower unit costs and in some cases the US Government offering huge industrial offsets. The Tornado ADV's nearest Western rivals, the F-15 and the Grumman (later Northrop Grumman) F-14 Tomcat were also limited in their export success. The only export sale for the F-14 was to Iran before the Islamic revolution overthrew the Shah in 1979. The F-15 achieved more with sales to Japan, Saudi Arabia and Israel. In the case of Israel, the US position was always stronger than any rivals due to the large multi-billion $ military aid packages Israel receives from the US. Politics also meant Japan was more likely to purchase US equipment and Saudi Arabia was coming into the US fold as the US increased its grip on the oil rich Middle East. Even on this side of the Atlantic BAe negotiated sales of the two-seat Hawk 100 advanced trainer and Hawk 200 single seat fighter to Oman at the expense of the contract for e8 Tornado F.3's mentioned above.

Chapter Five

Beyond the Millennium

A pair of Tornado F.3's from No.43(F) (background) and No.111(F) (foreground) squadrons during a flight in May 2004. The aircraft are armed with their post CSP armament of 4 AIM-120 AMRAAM and four ASRAAM representing probably the most formidable air to air armament in service on any fighter. Crown Copyright

By 1996, the Panavia Tornado ADV had been in RAF service for over 20 years, with 173 aircraft eventually being acquired for domestic use. Of this total, three aircraft were development prototypes and 18 aircraft were produced as Tornado F MK2's, while the remaining 152 were built to Tornado F MK3 standard. While the Tornado F.3 has proved to be an extremely competent interceptor for the RAF, its lack of agility and poor performance at medium and high altitudes has led to much criticism of both the aircraft and the government of the day for purchasing it. It is true to say that the Tornado F.3 is outclassed in the close combat arena by relatively cheap fighters such as the US F-16 Fighting Falcon and the Russian MiG-29 Fulcrum. However, the simple fact remains that the main role of the Tornado F.3 was long-range interception, not dog-fighting over the central front. The Tornado ADV was designed to fly long duration CAP's at distances of over 300 miles from base then engage and destroy enemy aircraft at as far a distance from the Tornado as possible. Smaller agile fighters in the F-16 and MiG-29 class were unable to compete with the Tornado F.3 in this arena.

Tornado F.3 ZG797 retaining No.29(F) squadron markings was used in support of the CSP, particularly for AIM-120 AMRAAM handling flight-testing. BAE Systems

This tactic would have worked well during the Cold War when the RAF had to counter the possibility of large-scale attacks on the United Kingdom by the then Soviet Union. During the 1991 Gulf War, the Tornado F.3 conducted the more suitable CAP role, although on a number of occasions aircraft were committed north into Kuwait and Iraq, particularly when the ground offensive began. However, in the post Cold War world the RAF has been tasked to participate in operations such as was seen in the skies of the former Yugoslavia in 1993-96 where air superiority over hostile territory was required. In this type of operation, the Tornado F.3 was finding itself at an increasing disadvantage. The introduction of active radar guided beyond visual range air to air missile such as the US AIM-120 AMRAAM and the Russian RVV-AE AA-12 'Adder' meant that the Tornado F.3 was also beginning to be outperformed by simpler cheaper fighters in the BVR arena from the early to mid-1990's. The Tornado F.3 armed with the semi-active Skyflash was considered to be being progressively outclassed in the BVR arena by aircraft armed with active radar guided missiles. Even a number of older designs like the McDonnell Douglas F-4F ICE (Improved Combat Efficiency) Phantom II in service with the German Luftwaffe and late model MiG-21 'Fishbeds' armed with modern radar and AMRAAM and 'Adder' respectively would be at certain advantages over Skyflash armed Tornado F.3's.

To be fair the Tornado F.3 performed well in the 1991 Gulf war. On the few occasions that contact was made with the enemy, the Iraqi aircraft turned north and headed for home when the Tornado's locked them up with their Foxhunter radar. The fact that USAF and Saudi F-15 Eagles racked up a large tally of air to air kills and no kills were achieved by either RSAF or RAF Tornado F.3s does not reflect on the performance of the Tornado. Most of the F-15 kills were achieved during offensive fighter sweeps over enemy territory, a role for which the Tornado F.3 was neither designed nor tasked. Following the 1991 war, the RAF developed tactics for use by the Tornado F.3 in the offensive fighter sweep or counter air role, exploiting the F.3's strengths and opponents weaknesses. During training exercises RAF Tornado F.3's have performed well against dedicated counter air fighters such as the USAF F-15C Eagle.

Top right: Tornado F.3 ZG797 is seen with wings in the maximum sweep position during flight testing to determine if there were any adverse handling qualities between the Tornado carriage aircraft and the AMRAAM missile. This successful series of flight test cleared the way for further integration flight-testing. BAE Systems

Above: The primary AMRAAM integration aircraft for the CSP was Tornado F.3 ZE155. This video still shows the AMRRAM ignition during the first Tornado launch of an AMRAAM on 16 December 1999. BAE Systems

A Tornado F.3 operating with the OEU flies in formation with ZH555 of No.5 squadron centre and ZH558 of No.43 squadron foreground. The OEU operated F.3 background is armed with ASRAAM and AMRAAM missiles, while the aircraft in No.5 squadron markings is carrying Skyflash and Sidewinder missiles. The No.43 squadron aircraft in the foreground is not configured with any external stores. Crown Copyright

RAF Tornado F.3's participating in operation Deny Flight over Bosnia Herzegovina were at a disadvantage as the rules of engagement meant that they had to visually identify their targets before engaging them. This could have brought the Tornado into the range of short-range air to air missiles carried by Serbian MiG-21 and MiG-29 fighters. To make matters worse, the main trade for the allied fighters were small helicopters, which frequently violated the No Fly Zone proving to be extremely hard targets for fighters to successfully intercept. At no time during the operation from 1993-1996 did RAF Tornado F.3's encounter or engage Serbian fixed wing fighter aircraft.

Originally, the RAF planned for a Tornado F.3 MLU (Mid Life Update) in the late 1980's. This was to run concurrently with the Tornado GR.1 MLU (which eventually brought RAF Tornado IDS up to GR MK4/4A standard). However, the F.3 MLU was eventually cancelled as the defence cuts of the late 1980's and early 1990's began to bite. The decision to begin replacement of RAF Tornado F.3 squadrons with Eurofighter before the RAF

Jaguar fleet contributed to the decision to abandon the MLU.

With increasing delays with the Eurofighter program, senior figures in the RAF expressed concern over the increasing delays in the introduction to operational service of the new fighter. To offset the shortfall in air defence capability, which was anticipated around the turn of the Century, the RAF wanted to proceed with a MLU program for up to 100 of its Tornado F.3 fleet. During winter 1995/96, the Ministry of Defence (MoD) conducted a review into the most cost-effective solution to equipping the RAF with a credible fighter aircraft to serve until the introduction of Eurofighter into full squadron service, then scheduled for 2005. As part of the review process, the MoD admitted in October 1995 that it was looking at various options including acquisition of surplus US McDonnell Douglas F/A-18C/D Hornets and Lockheed Martin F-16C/D Fighting Falcons. The MoD requested price and availability data on both types, but consistently denied reports that the defence secretary planned to drop the Tornado upgrade in favour of buying or leasing US fighters. In 1994, a MoD paper recommended that any Tornado F.3 upgrade program should be abandoned in favour of an off the shelf buy of surplus USAF F-16A/Bs. As the recommendation met with strong opposition from senior figures in the RAF the proposal was axed by the then defence secretary Malcolm Rifkind. Following this, it was then assumed that BAe would receive a contract valued at around £115 million to upgrade part of the Tornado F.3 fleet for service beyond the introduction of Eurofighter.

Prompting the emergence of the F-16 saga was the leaking of a short-sighted document drawn up by the special advisor to the defence secretary, claiming that upgrading the operational force of around 100 Tornado F.3s would be a waste of

By the early 2000's, Tornado F.3's modified under the CSP were firmly established in squadron service. However, the new advanced weapons were not incorporated until long after upgraded aircraft were available. ASRAAM entered service with Tornado F.3 squadrons in 2002. Here No.111(F) squadron's 80th anniversary painted aircraft ZE159 UV is seen carrying four ASRAAM missiles on the inboard wing station stub pylons. G H Lee

money. At the same time, it was reported that the Pentagon had offered to sell the RAF 40 new-build Lockheed Martin F-16C/D Block 50 aircraft for a unit price of $20m each or alternatively to lease them for $1m per year. As part of the offer it was also claimed that the US was trying again to interest the RAF in a potential buy of Lockheed Martin/Boeing F-22 (later re-designated F/A-22) Raptor fighters as an alternative to at least some of the planned RAF Eurofighter procurement. This immediately caused concern within the Eurofighter partner nations who were looking at the MoD for reassurance to the UK's commitment to the program. This reassurance was not long in coming, as almost as soon as the F-16 story broke it was categorically being denied by both the RAF and MoD. All through late 1995 and early 1996, a fierce row raged between pro-F-16 MP Mr Portillo

and Air Chief Marshall Sir Michael Graydon who was supported by the RAF Board and later the Deputy Prime Minister Michael Heseltine. They, along with BAe were rigorously pushing for the Tornado F.3 upgrade to go ahead. Mr Portillo eventually accepted the decision of the commons select committee on defence that recommended that the MoD spend £125 on upgrading the F.3.

In the 1970s, when the RAF was looking for a replacement for the McDonnell Douglas Phantom FGR.2/FG.1 and the remaining BAe Lightning's, the F-16 was one of the off-the-shelf US aircraft then considered. However, the F-16 was dropped from the competition at an early stage as the single engine design found no favour within the RAF and the F-16's AN/APG-66 radar did not meet the detection range capability required for the RAF air defence role.

Aircraft defending UK airspace were required to fly long CAP's (Combat Air Patrols) over the sea at ranges of some 300-400 miles from base, and then be capable of engaging enemy aircraft in the densest Electronic Counter Counter Measure (ECM) environment of any air combat scenario. For this reason the F-16 would still have been the wrong choice to meet this requirement as it lacks the endurance for the long range CAP of up to six

Three Tornado F.3's fly in formation for the camera. The farthest away aircraft is operated by the F.3 OEU testing ASRAAM and AMRAAM missile carriage on the aircraft. Crown Copyright

availability of Eurofighter. If the UK Government had not abandoned Mr Portillo's short-sighted proposal to procure F-16s in favour of the Tornado upgrade, then the potential harm this could have done to the Eurofighter and future European collaborative programs is incalculable.

The Go-ahead for the Tornado F.3 upgrade, which eventually became known as the CSP (Capability Sustainment Programme) was announced on 5 March 1996. The design and engineering work on the upgrade was to be undertaken by BAe Warton, while modification kits were to be produced by BAe Samlesbury. The contract called for the upgrade of 24 aircraft by BAe (now BAE Systems) personnel at RAF St Athan. The remaining 76 aircraft would be modified by combined RAF and DARA (Defence Aviation and Repair Agency) personnel at RAF St Athan using BAe supplied kits. Plans called for two squadrons to be equipped with upgraded aircraft by late 1998, with the remaining squadrons following later. The aircraft were then scheduled to remain in operational service until 2010. While the Tornado GR.1 aircraft which went through the MLU received knew designations of GR.4/4A, the Tornado F.3 CSP modified aircraft retained their F.3 designation, rather than adopt a new F.5 designation.

hours that is required for this type of operation. While the Northrop Grumman (formerly Westinghouse) AN/APG-68 radar of the current F-16C/D Block 50 was a capable lightweight multi-role radar, it failed to meet the detection range and capability of operating in the previously mentioned ECCM environment that the AI.24 Foxhunter radar was designed for. Another negative point of an F-16 purchase was that it was not compatible with the hose and drogue equipped RAF tanker force, instead relying on the fixed boom refuelling method. In addition, the RAF would have had to invest large sums in the extensive support infrastructure required for any F-16 acquisition. This, on-top of the estimated $800m sale price for a force of only 40 F-16C/Ds, not including spares support or the cost of training RAF air and ground crews to operate and support the F-16s would appear to have all contributed to the decision to upgrade the Tornado F.3. It was also claimed that the F-16 would cost some £600 per hour more to fly than the Tornado. According to the MoD "the £125m upgrade is a more modest investment representing better value for money."

In 1993, the United Kingdom convinced the Italian Air Force to abandon acquisition of surplus USAF F-15's or F-16's and instead adopt the Tornado F.3 as its stop gap fighter pending the

The main features of the upgrade involved modifications to the BAE Systems AI.24 Foxhunter radar and the incorporation of a MIL-STD-1553B digital data bus giving it ASRAAM and AIM-120 AMRAAM air to air missile capability - four of each would be carried. The radar upgrade adds a new processor giving automatic target acquisition, tracking and discrimination between head on targets by analysis of the radar signature of their first and second stage compressor blades, and the ability to conduct simultaneous multi-target engagements using AMRAAM.

Tornado F.3 squadrons operating with the Leuchars wing were the first to be declared operation with AMRAAM in May 2004. Here a No.43(F) squadron aircraft presents its underside revealing its complement of four AMRAAM missiles on the fuselage stations and four ASRAAM on the inboard wing stub pylons. Crown Copyright

The original F.3 MLU, which was proposed in the late 1980's included installation of the new Eurojet EJ200 turbofan being developed for Eurofighter, but designed with possible Tornado retrofits in mind. The dry thrust of an EJ200 is almost equivalent to the afterburner thrust of an RB199 MK104, 13,490-lb to 9,656-lb and 20,230-lb to 16,902-lb with afterburner. This increase in power would eliminate one of the Tornado F.3's main weaknesses of insufficient power for combat manoeuvring, particularly improving performance at medium and high altitude. Unfortunately for the RAF, the CSP did not involve installation of the EJ200. This was probably the right choice as the engine replacement program would have been extremely expensive and hard to justify with the Eurofighter preparing for production.

Part of the Tornado F.3 CSP improvements include enhancement of the main computer performance with a Power PC, an upgraded missile management system using ADA language, main computer software changes, new multi-function colour displays and the above mentioned duel redundant 1553B digital data bus. A CSP for the AI.24 Foxhunter radar, upgrading it to Stage 2G standard was also conducted in 1998/99. A Link 16 JTIDS terminal is also installed allowing greater situational awareness for the Tornado crew and increasing the versatility of the aircraft by allowing the receipt and transfer of information in near real time. In June 2002, BAE Systems was awarded an 8-year 'power by the hour' £75 million contract for support of the AI.24 Foxhunter until it's planned out of service date of 2010. The contract was the third Tornado support contract awarded following efforts of the Tornado Tiger Team. The first contract covered support for Tornado GR.4 avionics and the second covered repair of major structural items on Tornado airframes. Structural upgrades involving the partner companies and Turbounion are aimed it providing Tornado GR.4 airframes with an 8,000-hour flying lifetime.

Now able to exploit the full potential of the CSP/COV upgrades, a pair of Tornado F.3's, ZE207 GC from No.43(F) squadron (foreground) and ZE738 HC from N.111(F) squadron (background) fly in formation with a full load of four each of AMRAAM and ASRAAM. Crown Copyright

Simultaneously with the CSP, the Tornado F.3 fleet underwent the COV (Common Operational Value) upgrade program, which was basically an attempt to bring the fleet up to a common operational standard. Modifications were incorporated at RAF Coningsby and RAF St Athan when aircraft were due for scheduled major overhauls. Among the COV enhancements was activation of the defunct 400-kg (880-lb) fin fuel tank, improvements to the aircraft cooling system, enhancements to the HOTAS controls and improving functionality and lighting compatible with Night Vision Goggles.

As both the CSP and the COV programs called for deep strip down of the aircraft, it made sense to combine the programs, which would be completed simultaneously. From the 41st Tornado F.3 to be modified under the CSP the aircraft underwent the COV modifications simultaneously with both programs apparently being merged under the designation ADV2000.

The Tornado F.3 fleet has also undergone extensive structural enhancements allowing the fleet to continue towards its 2010 out of service date. Most aircraft went through the 25 FI (Fatigue Index) program which was designed to allow the F.3's to continue flying for an additional 37.5 FI. Beyond the 62.5 FI, the Tornado F.3's would require a new structural enhancement program to be undertaken. The Tornado F.3 fleet is already undergoing a MLFP (Mid-Life Fatigue Program). This included 24 ADV centre fuselage sections being structurally enhanced at EADS Aerostructures Augsburg plant in Germany under a contract awarded for the overhaul work in 2000. On 16 January 2003, EADS delivered the first renovated centre fuselage to the RAF. All 24-centre sections are planned for delivery by 2007. Some components including the wing boxes will be newly manufactured due to the high strain on the connection point of the variable-geometry wing. The wing boxes consist of milled titanium parts, which will be joined together using electron beam welding.

ZE159 from No.111(F) squadron shows the sleek lines of the Tornado F.3 with wings in the fully swept position. The aircraft is carrying a load of four ASRAAM's. G H Lee

As well as the CSP modifications, the effectiveness of the RAF Tornado F.3 fleet would be further increased when equipment programs, which were under trial in the mid-1990, were released for squadron service. These programs included the installation of Bofors BOL-304 flare dispensers in the rear of the Sidewinder launch rails. The second development Tornado ADV, A02 (ZA267) was involved in BOL-Sidewinder firing trials at Bascombe Down during 1995. Other measures to increase the self-defence capabilities of the Tornado F.3 include the acquisition of a variant of the GEC Marconi (now BAE Systems) Ariel towed radar decoy, which increased protection against radar guided missile threats. The system is housed in a modified Boz chaff/flare dispenser pod. Raytheon Systems Limited was selected as the supplier of the SIFF (Successor Identification Friend or Foe) for the Tornado F.3. In August 2000, Raytheon was also selected as SIFF supplier to be incorporated into a number of military systems with all three branches of the UK armed forces, including surface vessels, submarines, aircraft and missile systems. The SIFF includes interrogators, transponders, associated cryptographic computers and control panels as well as complete logistic support. The Tornado F.3 has a Have Quick secure radio, which complements the single U/VHF radio and a HF system. A Honeywell 764GT LINS (Laser Inertial Navigation System) with integrated GPS (Global Positioning System) further enhances the aircraft's navigational capability.

BAE Systems used a Tornado F.2 (ZD889) and a pair of Tornado F.3's (ZE155 and ZG797) in support of the CSP. The Tornado F.2 was ostensibly used as a radar target during developments of the Foxhunter Stage 2G radar upgrade. Resident Warton based ZE155 was

A pair of Tornado F.3's from RAF Leuchars sits on the wing of a Tristar in-flight refuelling tanker during a deployment to Canada in early summer 2004. Crown Copyright

the main development aircraft used to conduct AMRAAM test firings and eventually converted to more or less full CSP standard. ZG797 was used for AMRAAM handling characteristic trials and display software trials.

The CSP upgrade was centred on the ability to carry and launch the Raytheon AIM-120 AMRAAM and MBDA ASRAAM. Both these systems are described in chapter two. In mid-July 1998, captive carry flights began with a Tornado F.3 carrying AIM-120 AMRAAM missiles on the under fuselage stations. The flight tests were conducted from BAe Warton. The first 30-minute flight showed that there were no adverse effects on the aircraft's handling while carrying the missile clearing the way for further carriage tests and missiles launches, the first of which was conducted on 16 December 1999. Tornado F.3 ZE155 launched the AMRAAM over the Aberporth test range in the UK. The launch involved a SIV (Separation Integration Vehicle) missile, which was launched from a fuselage station. The missile

launch was conducted to evaluate the Tornado/AMRAAM separation characteristics. After launch the AMRAAM SIV flew a pre-programmed manoeuvre. The test launch was claimed as a complete success by the integration team paving the way for integration of the missile with the operational fleet. The RAF is using some of the 330 or so AMRAAM's purchased for the Royal Navy BAE Systems Sea Harrier FA.2 fighters. With the Sea Harrier being retired in 2006, the RAF will take over the complete AMRAAM stocks for use with the Tornado F.3 CSP aircraft and the Eurofighter Typhoon.

The original planned AMRAAM integration omitted the capability for mid-course guidance of the missile after launch from the Tornado. This was deemed unnecessary as Tornado tactics well practised during the Skyflash/Sidewinder era called for the launch of an infrared guided missile at the target as it manoeuvred to avoid the initial radar guided BVR missile shot. The poor performance of AMRAAM during longer duration flights when speed drops to as low as Mach 1 has allowed target aircraft to simply outrun the missile. This may have contributed to the deletion of mid-course guidance from the original AMRAAM/Tornado integration. However, the enhanced capability of the data-link

The MBDA ASRAAM has replaced the AIM-9L Sidewinder in RAF Tornado F.3 squadrons. The weapon entered service in 2002 and has been deployed operationally in the Persian Gulf and is also used for aircraft performing QRA alert duties in the United Kingdom. Author

updates to AMRAAM missile in flight to provide mid-course guidance was finally recognised when an AOP (AMRAAM Optimisation Program) contract was signed on 8 June 2001. This resulted in the Tornado F.3 being able to utilise the mid-course guidance capability of the missile allowing more accurate engagements at long-range. However, as pointed out before, the relatively poor performance of AMRAAM at longer ranges mean that the RAF's tried and tested tactics evolved with Skyflash and Sidewinder will still have a major role to play in the AMRAAM/ASRAAM era. AMRAAM was finally declared fully operational on Tornado F.3's of the Leuchars wing in May 2004.

The integration of ASRAAM has given the RAF Tornado F.3 probably the best infrared guided air to air missile capability anywhere in the world, although this would of course be disputed depending on what type of capability is deemed most important. ASRAAM may not be as manoeuvrable immediately off the launch rail as some modern short-range missiles like the Rafael Python IV. However, ASRAAM has a longer range and engagement envelope than most missiles in its

class with only the MBDA Mica infra red guided variant having a longer reach among modern missiles. This capability allows the Tornado to launch its weapon first then break-off or take evasive action to avoid any counter fire from the enemy aircraft. The term first shot first kill is often applied to more modern aircraft; however, the Tornado ASRAAM combination bestows to a certain degree this capability on the CSP modified aircraft in the infrared guided missile arena.

The downside to the Tornado/ASRAAM integration is the lack of a HMSS (Helmet Mounted Sighting System) on the Tornado, which means that the full high off-boresight or lock-on after launch capability of ASRAAM cannot be exploited. This has led to criticism of the integration with claims that it was a waste of time, money and is no real leap in capability over the AIM-9L/M Sidewinder. Of course, these claims are simply ludicrous and can only come from a complete lack of understanding of the ASRAAM's capabilities. Even in its reversionary analogue mode used on the Tornado F.3, ASRAAM bestows a quantum leap in capability over the archaic AIM-9 Sidewinder. The missile is much more manoeuvrable, faster, has a much longer range, more powerful warhead and more capable IIR (Imaging infrared) seeker head. ASRAAM also bestows upon the launch aircraft a Pseudo IRST (infrared Search and Track) system, which was lacking on the Tornado F.3.

A No.11 squadron Tornado F.3 carrying a pair of ALARM 2 missiles on the fuselage stations and four ASRAAM's on the inboard wing station stub-pylons.
G H Lee

There has been much criticism of the failure to integrate a HMSS on the Tornado F.3 to exploit the full capability of the weapon. The cost savings at tens on millions of £ may well be small in defence spending terms. However, with a mere 5 and a half years of service left, these costs would be harder to justify in a country which is already seeing billions of £'s of additional money going into unpopular, unwanted (by the public) and unauthorised (by the United Nations) conflicts to impose the US and UK's will on poorer nations. The fact that these same nations have had only a limited ability to defend themselves will undoubtedly lead to the conclusion that such systems are a luxury, which cannot be funded at the expense of civil programs.

Although there have been reports that the there were plans for integration of a HMSS on the Tornado F.3 even before it entered service, this was not part of the RAF original specification. Even, the BAe ACA (Agile Combat Aircraft) proposal of 1982, (forebear of the Eurofighter Typhoon) was not originally planned with a HMSS due to the lack of maturity of such systems in the early 1980's.

With funding for the Tornado F.3's replacement, Eurofighter Typhoon becoming problematic in 2004, diversion of further funds for exotic niceties

like a HMSS for the F.3 is unlikely in the current climate, even though experience of Tornado ADV/HMSS operations was gained using the Tornado F.2 TIARA. While a HMSS undoubtedly would enhance the Tornado F.3's close combat capabilities, the lack of such a system was unlikely to cause any serious problems for the F.3 force before it reached it retired in 2011.

The Tornado F.3 has always had the potential to be developed into a true MRCA with both air-to-air and air to surface capabilities. The AI.24 Foxhunter radar, although developed as a long-range air to air sensor for the RAF's Tornado ADV, always possessed the potential for use in the air to surface role. BAE Systems own documented records show that when the ACA was unveiled in 1982 as a future multi-role combat aircraft, its primary sensor was to be a variant of the Foxhunter augmented by a LRF (Laser Range Finder) for air to surface ranging and a FLIR (Forward Looking Infra-Red). The Foxhunter would have received a number of air to surface modes turning it into a true multi-role radar system.

In the late 1980's and early 1990's a number of advanced Tornado proposals were being put forward as a future strike aircraft for the RAF. The Tornado 2000 proposal, which emerged around 1991, was a proposal for an advanced strike aircraft optimised for low-level penetration of enemy air defences. This design appeared to have more in common with the Tornado ADV than the IDS variant. A fuselage stretch would have provided space for additional avionics and fuel, which together with a large semi-conformal fuel tank beneath the fuselage would have increased range by 25% over the Tornado IDS. The aircraft featured a faceted nose to reduce RCS (Radar Cross Section) and reduced RCS air intakes. Although there were reports of a proposed variant featuring a highly swept delta wing, the only artworks released by BAE showed the aircraft retaining the variable-geometry wing of the ADV/ECR/IDS. It must be assumed that an advanced multi-role radar system would have been incorporated such as the Blue Vixen developed for the Sea Harrier FA.2 or the ECR 90 (later Captor) developed for the Eurofighter Typhoon. The Tornado 2000 never got beyond the drawing board as the RAF was moving away from low-level operations to medium level operations.

BAe released this artwork of the proposed Tornado 2000 in the early 1990's, however, the Tornado 2000 never progressed beyond the proposal stage. BAE Systems

In 2003, a number of Tornado F.3's were operating with No.11 squadron at Leeming armed with the MBDA ALARM 2 (Air Launched Anti-Radiation Missile) for the lethal SEAD (Suppression of Enemy air Defences) role. Plans to introduce a SEAD capability on the Tornado F.3 have been a start stop affair since the Kosovo air campaign in 1999. With severe gaps in the capability of NATO capabilities for the suppression of enemy air defences identified during the Kosovo campaign, an UOR (Urgent Operational Requirement) was issued to equip the Tornado F.3's for the non-lethal SEAD role. The F.3's wing and tail glove mounted antenna for the RHWR had the potential to be used as an extremely capably ELS (Emitter Locator System) capable of identifying and locating air defence systems.

It is unclear whether or not the plans reached the flight test stage with the Tornado F.3 OEU; however, with the Kosovo conflict over the UOR was dropped. In 2002, with the probability of the Blair Labour government following the US and taking the UK into yet another unpopular and for the majority of the population an unjustified war, another UOR was issued to provide the Tornado F.3 with SEAD capability. Of course, with RAF and US aircraft having been conducting a low-key bombing campaign over Iraq since late 1998, it has to be wondered what exactly the sudden urgency was? While the bombing campaign achieved poor results against Iraq's air defence assets, these same completely obsolete air defence systems were unable to successfully counter the modern US and UK air power arrayed against it. At no point during the December 1998 to March 2003 campaign was Iraq able to shoot down any coalition aircraft with the exception of slow flying and easy to target uninhabited reconnaissance drones.

Qinitiq at Bascombe Down converted a Tornado F.3 for the new lethal SEAD role in 2002. It is though that around 12 aircraft and possibly a little more were converted for use by No.11 squadron based at Leeming. In early 2003, the first converted aircraft ZE763 DG was delivered to RAF Leeming. The aircraft could only carry ALARM on the fuselage stations with no capability to carry Skyflash or AMRAAM at the same time.

No.11 squadron did not deploy to support operation Telic (the UK code name for the invasion of Iraq) in March 2003. Although there have been reports that the squadron was able to deploy, it is doubtful that full training on the ALARM 2 had been completed. In addition, the poor state that Iraq's air defences were in prior to the invasion meant that available assets were deemed more than necessary to meet the extremely limited threat. However, possibly the main reason the ALARM equipped Tornado F.3's did not deploy was that the RAF F.3 detachment in the Gulf was operating from Saudi Arabia. The Saudi rulers did not allow coalition aircraft bombing Iraq to operate from Saudi territory. The F.3 detachment at Prince Sultan Air Base was allowed to operate because it was used for air defence and not offensive air to surface operations. This would have meant the ALARM equipped Tornado F.3's would have had to operate from an already overcrowded Kuwait air base requiring further support infrastructure to support the aircraft as the only Tornado variant to be operated from bases in Kuwait were the Tornado GR.4/A.

Top: The British Aerospace Experimental Aircraft Program (EAP) demonstrator aircraft, which first flew in August 1986, was a development stepping stone to today's Eurofighter Typhoon. This aircraft, used to flight test technologies for the then EFA (European Fighter Aircraft), now Eurofighter, utilised a modified Tornado rear section and vertical tail in its construction and was powered by the F.3's RB.199 Mk.104 engines. Above: An 11 Sqn Typhoon F.2 flies in formation with a Tornado F.3 in 11 Sqn colours.

The improvements to the F.3 achieved through upgrade programs enabled it to remain amongst the best long-range air defence interceptor aircraft in the world until retired in 2011. Crown Copyright

The main requirement for a QRA operation during and in the decade following the end of the Cold War was to intercept and investigate aircraft approaching the UK. Main trade came from Soviet and later Russian maritime patrol aircraft and bombers probing the UK air defence alert time. Following the terrorist attack on the Twin Towers of the World Trade Centre in New York City and the Pentagon in Washington on 11 September 2001, using hijacked airliners, which were flown into their targets with catastrophic effects, the QRA has also had the responsibility for possible interceptions of civil aircraft being used in possible terrorist attacks on targets in the UK. Although all manner of possible targets have been mentioned including state affairs, these are at best unlikely as Al Qaida's stated aim is to destroy the economic power of the west. Target sets have mainly been financial centres such as the Twin Towers and the HSBC bank. Tourist centres have also been attacked doing severe financial harm to that industry. Of course, the real horror of these attacks is the cost in lives.

Aircraft scrambled to investigate suspicious aircraft could use a number of measures to try and force the aircraft to land if suspicions remain after investigation. However, it is unclear whether there would be a mandate to shoot down a civil airliner, which did not respond to commands to land. Such an order would surely have to come from the Prime Minister, who would be dammed if he gave such an order and dammed if he did not give the order and the aircraft was used in a suicide attack. It should be remembered that before and since 11 September 2001, airliners were frequent targets for terrorist hijackings, many of which were resolved peacefully. It is important, therefore, that any such incident involving an aircraft is not immediately assumed to be on a suicide-bombing mission and shot down with the resultant murder of the passengers and crew. In reality, there is no real way on knowing until the aircraft either lands or is used as a weapon. It is highly unlikely in any event that interceptors could be scrambled and reach the target aircraft in time. It should be remembered that the attacks on 11 September 2001 were conducted within around 20 minutes from take-off at airports near the targets. There has been much talk of fighters being used to protect sites such as nuclear power stations. Ironically, British and US military

The Sun finally set on the Tornado F.3 in RAF service in March 2011. BAE Systems/G H Lee

aircraft have proved to be the main threat to these same power stations with a number of unauthorised over flights of the stations by US and British combat aircraft. At no time does there appear to have been any attempt to intercept or investigate these incidents emphasising the difficulties in combating a threat from within the UK's borders. Facilities were provided at a number of other RAF Stations around the country to provide limited QRA facilities for Tornado F.3 detachments in times of tension. However, terrorist attacks by their very nature are not pre-advertised allowing the luxury of pre-positioning military equipment to counter them.

The August 2004 RAF complement of four operational squadrons and OCU (Reserve squadron) equipped with the Tornado F.3 called for some 87 aircraft to be available for operations. The reduction of one Tornado F.3 squadron announced in July 2004 saw the required number of available aircraft drop to around 72 airframes with others held in reserve. The reduction in the required aircraft numbers, combined with the MLFP allowed the RAF to continue to operate the F.3 until its final retirement, by spreading flight hours over a larger number of airframes, particularly once the ALARM role was relinquished.

As the Tornado F.3 drawdown gathered pace 25 Squadron was disbanded at Leeming on 4 April 2008 and the force concentrated at RAF Leuchars, with 43 and 111 Squadrons providing the Northern UK QRA as well as providing aircraft and crews for the Tornado F.3 detachment at RAF Mount Pleasant in the Falkland Islands. No.56(R) Squadron, the F.3 OCU, which had moved to RAF Leuchars in late March 2003, officially merged with No.43 Squadron on 25 April 2008. The merged unit retained the 43 Squadron number plate, but took on the role of Tornado F.3 crew conversion as well as other course training such as the Qualified Weapon Instructor Course. For its dual role of operational squadron and training unit, the squadron had a significantly larger complement of aircraft and personnel than a standard operational squadron, with an average of 26 aircraft and 30 crews.

In the final drawdown, No.43 Squadron disbanded in July 2009 and in September that year, Tranche 2 Typhoon FGR.4's replaced the Tornado F.3's of 1435 Flight in the Falkland Islands. The last Tornado F.3 squadron in RAF service, No.111 Squadron, disbanded on 22 March 2011, bringing to an end the operational career of the aircraft in RAF service.

Four aircraft continued flying with QinetiQ on development of the MBDA Meteor AAM until they were retired following the last development flight on 20 June 2012.

Appendices

Appendix I

Specification Tornado F MK3

Powerplant: Two x Turbounion RB199 MK104E afterburning turbofan engines, each rated at 42.9-kN (9,656-lb) dry and 75.2-kN (16,410-lb) in afterburner.
Wingspan: 8.6-m (28-ft 2-in) swept in 67-degree position and 13.91-m (45-ft 7.5-in) in un-swept position.
Wing area: 30.00-m2 (322.93-sq ft) in 25 degree position.
Wing sweep: 25 to 67 degree
Length:
Height: 5.95-m (19-ft 6-in)
Empty weight: 14500-kg (31,967-lb
Maximum take-off weight: 28000-kg (61,729-lb)
Internal fuel: 4700-kg in fuselage and wing fuel tanks and 440-kg in vertical tail fuel tank.
External fuel: Two x 1800-kg external fuel tanks
Maximum speed: Mach 2.2 (1,452 mph/2337 km/h) at altitude, Mach 1.2 (800-kts/921 mph) at see level.
Service Ceiling: 15250-m+ (50,000-ft+)
Range: 1853-KM (1,151 miles) on a subsonic interception or 556-km (354 miles) on a supersonic interception.
Take-off run: 762-m (2,500-ft)
Landing roll: 370-m (1,215-ft)
G-limits: +7.5
Armament: Four MBDA Skyflash semi-active radar homing or AIM-120B active radar guided AMRAAM beyond visual range air to air missiles in staggered recesses under the fuselage and four AIM-9L/M or MBDA ASRAAM infrared guided air to air missiles on the inboard wind station stub pylons. A single 27-mm Mauser cannon in located on the starboard forward fuselage.

General Description

Tornado ADV (Air Defence Variant), designated F3 by the Royal Air Force, has been developed to fulfil a United Kingdom requirement for a long-range, long-endurance, fighter/interceptor aircraft. It is in service with both the Royal Air Force and the Royal Saudi Air Force.

Armed with four Sidewinder short-range air-to-air missiles, four Sky Flash medium range air-to-air missiles, and capable of in-flight refuelling, Tornado ADV can maintain combat air patrol for many hours. The long range search and track capability of its Foxhunter air intercept radar enables targets to be successfully engaged at Beyond-Visual-Range (BVR).

Dimensions

Wing span (unswept)	13.9 m	45 ft 7 in
Wing span (swept)	8.6 m	28 ft 2 in
Length	18.6 m	61 ft 0 in
Height	6.0 m	19 ft 8 in

Weights and Fuel

Max Take-Off-Weight (MTOW)	28,000 kg	61,700 lb
Empty weight	14,450 kg	31,800 lb
Max internal fuel	7,270 Litres	1,600 Imp Gal
Max external fuel	7,500 Litres	1,650 Imp Gal

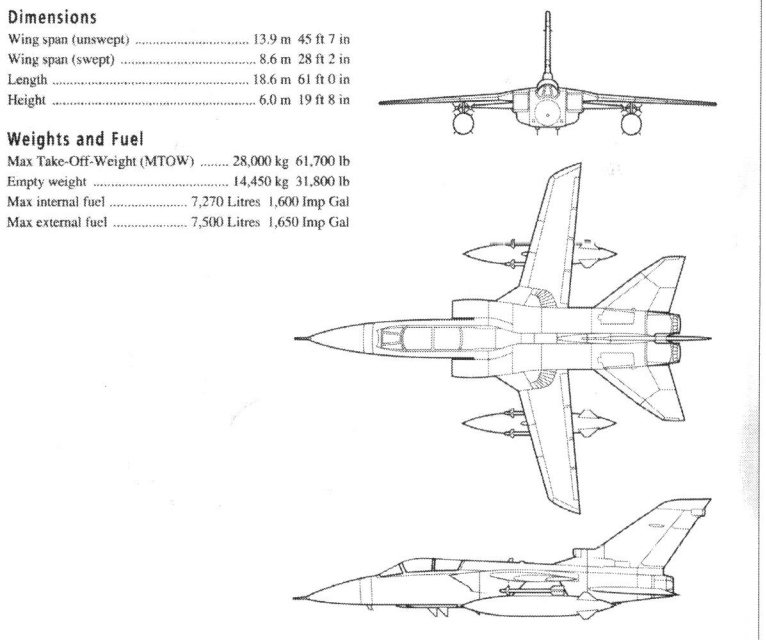

PANAVIA
Arbellastrasse 16,
8. München 86.
Germany

BRITISH AEROSPACE PLC
Warwick House.
Farnborough Aerospace Centre.
England

ALENIA
P. le Tecchio 51/A,
Napoli.
Italy

DASA
8 München 80.
Postfach 801109.
Germany

Appendix II

Tornado ADV Variants

Tornado F MK2: Three, prototypes and 18 production aircraft delivered with RB199 MK103 engines.
Tornado F MK2A: Planned upgrade of 18 production Tornado F MK2's almost to F MK3 standard. Never undertaken.
Tornado F MK3: 152 standard production aircraft with up-rated RB199 MK104 engine additional weapon carrying capability and auto-wing sweep built for the RAF. Twenty-four F MK3's built for Saudi Arabia.
Tornado EF MK3: Designation applied to between 12 and 14 Tornado F MK3's converted in early 2003 to carry the MBDA ALARM 2 missile for the suppression of enemy air defence role. The designation is thought to be temporary as the aircraft converted under an Urgent Operational Requirement may be returned to F MK3 standard unless a release to service is obtained.

The RAF had an initial requirement for 165 Tornado ADV, of which the first 3 aircraft were to be prototypes added to Batch One Tornado production. Of the Batch 6 aircraft 24 were diverted to fulfil an order from the RSAF, being replaced from Batch 7 production. The RAF ordered a further 15 aircraft to be delivered from Batch 8, but 7 of these were subsequently cancelled under the 1990 options for change defence cuts. The remaining 8 originally ordered for Oman were delivered from Batch 8 with the RAF taking delivery of its last Tornado F MK3 in March 1993. As there were differences particularly in the standard of radar software between aircraft of the same production Batch Block numbers become more important. The 18 aircraft from Batch 4 production were the only ADV completed as Tornado F MK2'.

Appendix III

Batch/Block	Standard	Dual control	Total	First Flight
Batch 1 Block 1	2	1	3	27.10.1979
Batch 4 Block 8	0	6	6	12.04.1984
Batch 4 Block 9	10	2	12	11.01.1985
Batch 5 Block 10	12	6	18	20.11.1985
Batch 5 Block 11	22	12	34	26.08.1986
Batch 6 Block 12	39	7	46	09.1987
Batch 6 Block 13	12	10	22	01.12.1988
Batch 7 Block 14	7	0	7	1990
Batch 7 Block 15	17	0	17	1991
Batch 8 Block 16	0	8	8	02.1992
TOTALS	121	52	173	

Appendix IV

RAF Tornado ADV Serial numbers

Batch 1: ZA254, ZA267, ZA283
Batch 4: ZD899-ZD906, ZD932-ZD941
Batch 5: ZE154(T), ZE155-ZE168, ZE199-ZE210, ZE250-ZE258, XE287-ZE296, ZE338-ZE343
Batch 6: ZE728-ZE737, ZE755-ZE764, ZE785-ZE794, ZE808-ZE812, ZE830-ZE839, ZE858, ZE862, ZE887-ZE889, ZE907-ZE908, ZE911, ZE934, ZE936, ZE941-ZE942, ZE96-ZE969, ZE982-ZE983,
Batch 7: ZG728, ZG730-ZG735, ZG751, ZG753, ZG755, ZG757, ZG768, ZG770, ZG772, ZG774, ZG776, ZG778, ZH780, ZG793, ZG795-ZG799
Batch 8: ZH552-ZH559
Aircraft serials ZA267, ZD899-904, ZD934-935, ZE154, ZE157, ZE160, ZE163, ZE166, ZE202, ZE205, ZE208, ZE256, ZE296, ZE340, ZE343, ZE728-729, ZE735, ZE759, ZE786, ZE791, ZE793-794, ZE810, ZE812, ZE832, ZE839, ZE858, ZE862, ZE934, ZE941 are confirmed twin-stick aircraft

Glossary

ACMI	Air Combat Manoeuvring Instrumentation
ADV	Air Defence Variant
AEW	Airborne Early Warning
AFVG	Anglo-French Variable Geometry
AGM	Air to Ground Missile
AI	Airborne Interception
AIM	Airborne Interception Missile
ALARM	Air Launched Anti-Radiation Missile
AMLCD	Active Matrix Liquid Crystal Display
AMRAAM	Advanced Medium Range Air to Air Missile
ASR	Air Staff Requirement
ASRAAM	Advanced Short Range Air to Air Missile
AST	Air Staff Target
AVS	Advanced Vertical Strike
B	Bomber
BAC	British Aircraft Corporation
BAE	British Aerospace
BITE	Built In Test Equipment
BVR	Beyond Visual Range
COV	Common Operational Value
CRT	Cathoyde Ray Tube
CSP	Capability Sustainment Program
DARA	Defence Aviation Repair Agency
DASA	
E	Electronic
EADS	
ECR	Electronic Combat and Reconnaissance
ECCM	Electronic Counter Counter Measures
ECM	Electronic Counter Measures
F	Fighter
FA	Fighter Attack
FAA	Fleet Air Arm
FADEC	Full Authority Digital Engine Control
FG	Fighter Ground attack
FGR	Fighter Ground attack and Reconnaissance
FLIR	Forward Looking Infra-Red
FMICW	Frequency Modulated Interrupted Continuous Wave
GCI	Ground Control Interception
GPS	Global Positioning System
GR	Ground attack and Reconnaissance
HARM	High Speed Anti-Radiation Missile
HAS	Hardened Aircraft Shelter
HOTAS	Hands-On Throttle And Stick
HP	High Pressure
HUD	Heads-Up Display
IDS	Interdictor Strike

INS	Inertial Navigation System
IP	Intermediate Pressure
IRST	Infra-Red Search and Track
JASDF	Japan Air Self Defence Force
JTIDS	Joint Tactical Information Distribution System
LINS	Laser Inertial Navigation System
LP	Low Pressure
MBDA	Matra BAe Dynamics Alenia
MBB	Messerscmitt-Bolkow-Blohm
MLFP	Mid-Life Fatigue Program
MoA	Memorandum of Agreement
MoU	Memorandum of Understanding
MRAAM	Medium Range Air to Air Missile
MRCA	Multi-Role Combat Aircraft
MTU	
NAMMA	NATO Multi-role combat aircraft Management Agency
NATO	North Atlantic Treaty Organisation
NETMA	NATO Eurofighter Tornado Management Agency
OCU	Operational Conversion Unit
OEU	Operational Evaluation Unit
QRA	Quick Reaction Alert
RAF	Royal Air Force
RAFO	Royal Air Force of Oman
RHWR	Radar Homing and Warning Receiver
RSAF	Royal Saudi Air Force
RWR	Radar Warning Receiver
SACEUR	Supreme Allied Command Europe
SACLANT	Supreme Allied Command Atlantic
SAM	Surface to Air Missile
SARH	Semi-Active Radar Homing
SEPECAT	
SOC	Sector Operations Centre
SRAAM	Short Range Air to Air Missile
T	Trainer
TIALD	Thermal Imaging Airborne Laser Designator
TIARA	Tornado Integrated Avionics Testbed
TDMA	
TNR	Tornado Nose Radar
TSR	Tactical Strike Reconnaissance
TV	Television
UHF	Ultra High Frequency
UK	United Kingdom
UOR	Urgent Operational Requirement
US	United States
USAF	United States Air Force
USN	United States Navy
VHF	Very High Frequency
V/STOL	Vertical/Short Take-Off and Landing
VTOL	Vertical Take-Off and Landing

Centurion Publishing

ISBN 10: 1-903630-38-X
ISBN 13: 978-1-903630-38-9

Printed in Great Britain
by Amazon

86532965R00086